SUSTAINABLE USE OF LAND AND WATER UNDER RAINFED AND DEFICIT IRRIGATION CONDITIONS IN OGUN-OSUN RIVER BASIN, NIGERIA

Omotayo Babawande Adeboye

Thesis committee

Promotor

Prof. Dr E. Schultz
Emeritus Professor of Land and Water Development
UNESCO-IHE Institute for Water Education
Delft, the Netherlands

Co-promotors

Prof. Dr K.O. Adekalu
Professor of Soil and Water Resources Engineering
Department of Agricultural and Environmental Engineering
Obafemi Awolowo University
Ile-Ife, Nigeria

Dr K. Prasad
Irrigation and Water Resources Management Consultant
Lalitpur, Nepal

Other members

Prof. Dr P.J.G.J. Hellegers, Wageningen University
Prof. F.A. Adeniji, University of Maiduguri, Nigeria
Dr Eyasu Yazew Hagos, Mekelle University, Ethiopia
Dr F.W.M. van Steenbergen, MetaMeta Research, 's Hertogenbosch

This research was conducted under the auspices of the SENSE Research School for Socio-Economic and Natural Sciences of the Environment.

SUSTAINABLE USE OF LAND AND WATER UNDER RAINFED AND DEFICIT IRRIGATION CONDITIONS IN OGUN-OSUN RIVER BASIN, NIGERIA

Thesis

Submitted in fulfilment of the requirements of
the Academic Board of Wageningen University and
the Academic Board of the UNESCO-IHE Institute for Water Education
for the degree of doctor
to be defended in public
on Thursday 28, May 2015 at 10:30 a.m.
in Delft, the Netherlands

by

Omotayo Babawande Adeboye
Born in Modakeke-Ife, Nigeria

RC Press/Balkema is an imprint of the Taylor & Francis Group, an informa business

© 2015, Omotayo Babawande Adeboye

Published by:
CRC Press/Balkema
PO Box 11320, 2301 EH Leiden, The Netherlands
e-mail: Pub.NL@taylorandfrancis.com
www.crcpress.com – www.taylorandfrancis.com

ISBN 978-1-138-02841-8 (Taylor & Francis Group)
ISBN 978-94-6257-278-2 (Wageningen University)

Table of contents

Page

Table of contents   v
List of figures   xi
List of tables   xv

Acknowledgements   xvii
Summary   xix

1   Introduction   1
      1.1   Structure of the thesis   3
      1.2   Scope of the thesis   4

2   Background and objectives   7
      2.1   Current challenges in water resources development and management   7
      2.2   Food security and irrigated agriculture   8
      2.3   Challenges of irrigated and rainfed agriculture   9
      2.4   General profile of Nigeria and geographic location   10
             2.4.1   Climate   10
             2.4.2   Administrative divisions   11
             2.4.3   Economy   12
      2.5   Water resources management and state of use in Nigeria   13
      2.6   Problem description   17
      2.7   Research questions   19
      2.8   Hypothesis   19
      2.9   Expected outcomes and contributions to knowledge   20
      2.10   Research objectives   20

3   Ogun-Osun River Basin, Nigeria   21
      3.1   Climate   21
      3.2   Rainfall pattern and air temperature   22
      3.3   Soil and land resources   23
      3.4   Current challenges in Ogun-Osun River Basin   24

4   Soybeans *(Glyxine Max(M.) Merr)*   27
      4.1   General botany and descriptions   27
      4.2   Growth and development   28
             4.2.1   Vegetative growth   28
             4.2.2   Reproductive growth and maturity   29
      4.3   Seed yield   29
      4.4   Origin and distribution   29
      4.5   Production practices in Nigeria   33
      4.6   Prospect for Soybean production in Nigeria   33

5   Crop yield models   35
      5.1   Previous crop yield models   35
             5.1.1   Group I models   35

		5.1.2 Group II models	37
		5.1.3 Group III models	40
	5.2	Computer based crop yield models: AquaCrop	41
		5.2.1 Introduction	41
		5.2.2 Description of components	42
		5.2.3 Water productivity and above ground biomass	46
		5.2.4 Responses to water stress	48
		5.2.5 Root extension and water extraction	52
		5.2.6 Responses of harvest index (HI) to water stress	53
		5.2.7 Criterion for the model evaluation	55
	5.3	Radiation use efficiency (RUE)	57
	5.4	Concept of irrigation water use efficiency and water productivity (WP)	59
	5.5	Simple model for carbon assimilation in plants	60
		5.5.1 Light based model	60
		5.5.2 Water based model	61

6		Research methodology	63
	6.1	Study area	63
	6.2	Experimental treatments, field lay out, cultivation and measurements in the rainy seasons	63
		6.2.1 Experimental treatments	63
		6.2.2 Field lay out, cultivation practices and measurements	64
		6.2.3 Leaf area index and fraction of Intercepted Photosynthetically Active Radiation	66
		6.2.4 Dry above ground biomass	67
		6.2.5. Radiation Use Efficiency	67
		6.2.6 Crop transpiration	68
		6.2.7 Actual crop evapotranspiration	69
		6.2.8 Field observations of plant phenologic development in the rainy season of 2011	70
		6.2.9 Field observations of plant phenologic development in the rainy season of 2012	70
	6.3	Experimental treatments, field lay out, cultivation and measurements in the dry seasons	72
		6.3.1 Experimental treatments	72
		6.3.2 Field lay out, cultivation and measurements in the dry seasons	72
		6.3.3 Measurement of soil water evaporation	76
		6.3.4 Actual crop evapotranspiration in the dry season	77
		6.3.5 Dry above ground biomass under water deficit conditions	77
		6.3.6 Field observations of plant phenologic development in the dry season of 2013	78
		6.3.7 Field observations of plant phenologic development in the dry season of 2013/2014	78
	6.4	Economic analysis	79
		6.4.1 Land limited conditions	79
		6.4.2 Water limited conditions	80
	6.5	Statistical analysis	80
		6.5.1 Rainfed conditions	80
		6.5.2 Irrigated conditions	80

7   Water conservation, soil water balance and productivity of Soybeans under
        rainfed conditions   81
        7.1   Environmental conditions in the rainy seasons   82
        7.2   Physical and chemical properties of the soil   84
        7.3   Moisture content and temperature in the soil   84
                7.3.1 Distribution of moisture in the soil   84
                7.3.2 Temperature in the soil   86
        7.4   Seasonal soil water storage (SWS)   88
        7.5   Biometric measurements and analysis of the plant canopy   89
                7.5.1 Plant height and number of leaves   89
                7.5.2 Extinction coefficient, Leaf area index and fraction of
                      Intercepted Photosynthetically Active Radiation   90
        7.6   Radiation Use Efficiency   98
        7.7   Use of light model in simulating dry matter   100
        7.8   Discussion   100
        7.9   Conclusion   103
        7.10   Soil water balance   104
                7.10.1 Seasonal evaporation from the soil and transpiration from the
                       plants   104
                7.10.2 Seasonal transpiration efficiency (TE) and RUE   107
                7.10.3 Soil water storage and dry above ground biomass at harvest   107
                7.10.4 Water productivity for biomass (WP$_{biomass}$), Intercepted
                       Photosynthetically Active Radiation (IPAR) and RUE   108
        7.11   Soil water storage and crop evapotranspiration (SWU)   108
        7.12   Soil water storage and seed yield   109
                7.12.1 Harvest index (HI), SWS and SWU   110
        7.13   Yield, water productivity, harvest index and RUE   110
                7.13.1 Seasonal crop water use and seed yield   110
                7.13.2 Seasonal crop water use and water productivity for seed
                       (WP$_{seed}$)   111
                7.13.3 Water productivity, seed yield and biomass at harvest   112
                7.13.4 Seed yield, IPAR and RUE   113
                7.13.5 Water productivity for seed, IPAR and RUE   113
                7.13.6 Harvest index, IPAR and RUE   114
        7.14   Partitioning of the dry above ground biomass at harvest   114
        7.15   Economic evaluation   115
                7.15.1 Land limiting conditions   115
                7.15.2 Water limiting conditions   116
        7.16   Discussion   116
        7.17   Conclusion   119

8   Effects of deficit irrigation on soil water balance, yield and water
        productivity of Soybeans   121
        8.1   Environmental conditions during the irrigation seasons   122
        8.2   Biometrics and growth parameters   125
                8.2.1 Leaf area index and dry matter   125
                8.2.2 Relationship between accumulation of dry matter, seed yield
                      and seasonal water use   129
                8.2.3 Relationship between yield decrease and decrease in
                      evapotranspiration   130

8.3   Soil water balance   132
        8.3.1 Seasonal soil evaporation (SEP$_i$), transpiration (STP$_i$) and
               water use (SWU$_i$)   132
8.4   Water productivity (WP) and irrigation water productivity (IWP)   134
8.5   Water productivity and harvest index   135
8.6   Effects of deficit irrigation on yield components   136
8.7   Economic evaluation   137
        8.7.1 Land limited conditions   137
        8.7.2 Water limited conditions   138
8.8   Conclusion   139

9   Modelling of response of the growth and yield of Soybeans to full and
        deficit irrigation by using the AquaCrop model   141
9.1   Introduction   141
9.2   Input data requirement of the AquaCrop model   143
        9.2.1 Environmental conditions   143
        9.2.2 Crop parameters   144
        9.2.3 Soil parameters   146
        9.2.4 Irrigation and field management   146
9.3   Calibration of the AquaCrop model   146
        9.3.1 Calibration of irrigation parameters   146
        9.3.2 Calibration of field management practices   147
        9.3.3 Calibration of crop parameters   148
9.4   Validation of the AquaCrop model   150
9.5   Criteria for evaluating the AquaCrop model   150
9.6   Water productivity   150
9.7   Results   150
        9.7.1 Canopy cover   150
        9.7.2 Dry above ground biomass   151
        9.7.3 Soil moisture content   152
9.8   Validation results of the AquaCrop model   153
9.9   Discussion   156
9.10   Conclusion and remarks   157

10   Evaluation   159

11   References   163

Appendices   185

A   List of symbols   185
B   Acronyms   189
C   Meteorological variables logged at intervals of 10 minutes at the
        experimental site from 2011 - 2014   191
D   Average soil temperatures in the upper 30 and lower 30 cm of the soil
        profile under different conservation practices and the conventional method
        in the rainy seasons   193
E   Analysis of variance (ANOVA) of soil water storage   195
        E 1   Statistical analysis of the soil water storage in 2011   195
        E.2   Statistical analysis of the soil water storage in 2012   197

E.3   Statistical analysis of the (a) plant heights and (b) number of leaves
       in 2011 under rainfed conditions   197
E.4   Statistical analysis of the (a) plant heights and (b) number of leaves
       in 2012 under rainfed conditions   199
F   Analysis of variance (ANOVA) of the leaf area indices in (a) 2011 and (b)
       2012 under rainfed conditions   203
G   Analysis of variance (ANOVA) fractions of PAR at different stages of the
       growth of the crop in (a) 2011 and (b) 2012   207
H   Partitioning of evapotranspiration into transpiration and evaporation for
       water conservation in 2012   211
I   Analysis of variance of the yield of the crop in (a) 2011 and (b) 2012   213
J   ANOVA of water productivity of the crop in (a) 2011 and (b) 2012   215
K   ANOVA of harvest indices of the crop in 2011 and 2012 under rainfed
       conditions   217
L   Cost analysis for the six water conservation and the conventional practices
       under rainfed conditions   219
M   ANOVA of LAIs in the dry season   221
N   ANOVA of dry matter in the dry season   223
O   ANOVA of seed yields in the dry season   225
P   Analysis of the cost of production under irrigation conditions   227
Q   Samenvatting   229
R   About the author   237

SENSE Diploma

List of Figures

Figure	Description	Page
1.1	Global freshwater use (Food and Agriculture Organization of the United Nations (FAO), 2010b)	2
2.1	Six Geo-political zones in Nigeria	12
2.2	Population trend in Nigeria since 1950 and the projections (medium variant) for the future (World Population Prospects (WPP), 2012)	13
2.3	River basins in Nigeria (FAO, 2005)	14
2.4	Water withdrawal in Nigeria (FAO, 2005)	15
2.5	Structure of the irrigation sector in Nigeria in 2004 (FAO, 2005)	15
2.6	Population distribution in Ogun-Osun River Basin, Nigeria (Federal Ministry of Environment, Nigeria (FMEN), 2010)	18
3.1	Ogun-Osun River Basin and the locations of other River Basins in Nigeria (adapted from Areola et al., 1985)	21
3.2	Land use pattern in Ogun-Osun River Basin, Nigeria	22
3.3	Mean annual rainfall in Ogun-Osun River Basin, Nigeria (FMEN, 2010)	23
3.4	Soil and vegetation patterns in Ogun-Osun River Basin (Ogun-Osun River Basin and Rural Development Authority (OORBA), 1982)	24
3.5	Severity of human induced land degradation in Nigeria (FAO, 2010c)	25
4.1	A young vegetative Soybean plant (Roth, 2013)	27
5.1	AquaCrop indicating the functional relationship between the components of the model in the soil-plant-atmosphere continuum and the parameters driving the phenology, canopy cover, transpiration, biomass production and final yield. Irrigation (I); Min air temperature (T_n); Max air temperature (T_x); Reference evapotranspiration (ET_o); Soil evaporation (E); Canopy transpiration (T_r); Stomatal conductance (gs); Water productivity (WP); Harvest index (HI); Atmospheric carbon dioxide (CO_2) concentration; (1), (2), (3), (4) represent different water stress response functions). Continuous lines indicate direct links between variables and processes. Dashed lines indicate feedbacks (Raes et al., 2012).	42
5.2	Schematic representation of canopy development during the exponential growth and the exponential decay stages (Raes et al., 2012)	49
5.3	Reduction in green canopy cover (CC) during senescence for various canopy decline coefficients (CDC). All curves have initial green (CC_o) at 0.9 and starting time at 0 (Raes et al., 2012)	50
5.4	Variation of green CC from emergence until physiological maturity under non-stress conditions (Raes et al., 2012)	51
5.5	Schematic representation of rooting depth along the crop growth cycle from sowing until maximum (shaded area) effective rooting depth (Raes et al., 2012)	53
6.1	Location of the experimental fields at the Teaching and Research Farms of Obafemi Awolowo University, Ile-Ife	63

6.2   Experimental fields showing the treatments and their lay out during
        the rainfed experiments in 2011 and 2012   65
6.3   Lay out of the experimental treatments during the rainy seasons   66
6.4   Experimental field during rainfed conditions showing: (a) ploughed
        land in preparation for the cultivation; (b) measurement of plant
        biometrics; (c and d) sampling of LAIs by using a Ceptometer; (e)
        Soybean during the seed filling stage; (f) Senescence at late stage; (g
        and h) late stage when about 70% of the pods were ripe and matured   71
6.5   Lay out of the treatments in the dry seasons   73
7.1   Variability in volumetric moisture content in the upper 60 cm of the
        soil profiles under rainfed conditions   86
7.2   Variability in volumetric moisture content among the treatments
        during establishment, flowering, pod initiation, seed filling and near
        maturity under rainfed conditions. Each dot represents the mean of
        the soil moisture from two replicates   87
7.3   Average temperatures in the upper 60 cm of the soil profiles under
        rainfed conditions. Each bar represents the mean of the soil
        temperatures from two replicates   88
7.4   Soil water storage at five growth stages of the plant during rainfed
        conditions. Each bar represents average soil water storage from two
        replicates. Bars with the same letter indicate that the soil water
        storages are not significantly different at $p < 0.05$ by using Duncan
        multiple comparison of means   89
7.5   Relationship between natural log of the intercepted PARs and leaf
        area indices of the treatments in the rainy seasons. The slopes of the
        regression lines are the extinction coefficients of the crop   92
7.6   Changes in the leaf area indices of Soybeans along phenologic stages
        in the rainy season. Each dot represents the mean with standard error
        of four replicates in the treatments   94
7.7   Leaf area indices and fIPAR at extinction coefficients of 0.46 and
        0.51 during the rainy seasons. Each dot represents the mean of LAIs
        from four replicates   96
7.8   Daily cumulative above ground biomass from emergence until
        maturity under rainfed conditions. Each dot represents the mean and
        standard error for four replicates   97
7.9   Radiation use efficiency of Soybean under rainfed condition by
        using (a) cumulative PARs derived from Global solar radiation and
        (b) PARs from instantaneous measurements (data after senescence of
        the plant at the late stage in each treatment were not included).
        Average of the dry above ground biomass for each day from four
        replicates was used to compute the RUE   100
7.10   Measured and simulated dry above ground biomass by using a
        simple assimilation model under rainfed conditions   101
7.11   Daily soil evaporation and transpiration estimated by using the single
        crop coefficient approach, fraction of the intercepted PAR radiation
        for the Soil bund (BD) under rainfed conditions   105
7.12   Production of (a) biomass in relation to a seasonal crop water use.
        The slope of the line represents $WP_{biomass}$ for Soybean in Ile-Ife. The
        intercept on the x-axis denotes the minimum amount of water lost by
        evaporation from the soil surface. Reduction in (b) transpiration
        efficiency with seasonal transpiration under rainfed conditions   106

7.13   Seasonal soil water storages and crop water use under rainfed
        conditions   109

7.14   Seasonal soil water storages and seed yields of Soybeans under
        rainfed conditions. Each bar is the average of soil water storage from
        two replicates and seed yields from four replicates   110

7.15   Partitioning of the dry above ground biomass into dry stems, chaffs,
        and seeds after harvest in the rainy seasons. Each bar represents the
        mean with standard error from four replicates   114

7.16   (a) Seed yield versus seasonal crop water use and (b) production cost
        versus seasonal water use in land limiting conditions   115

8.1   Application of water to Soybeans by using in-line drip irrigation   125

8.2   Relationship between LAIs during seed filling and (a) plant height;
        (b) number of leaves; (c) number of seeds per pod; (d) number of
        pods per plant stand in both irrigation seasons   126

8.3   Relationships between dry above ground biomass at harvest and (a)
        plant height at harvest; (b) seasonal transpiration; (c) number pods
        per plant; (d) number of seeds per pod in both irrigation seasons.   128

8.4   Seasonal crop water use, seed yield and accumulation of dry matter
        in the dry seasons   130

8.5   Relationship between the decrease in relative yield and deficit in
        seasonal relative evapotranspiration for the two irrigation seasons   131

8.6   Relationship between LAIs during seed filling and (a) seed yield; (b)
        STP$_i$ in the two seasons   134

8.7   Water productivity as a function of harvest index of TGX 1448 2E
        Soybeans grown under drip irrigated conditions with five treatments   136

8.8   Components of the yield at harvest in both irrigation seasons. Each
        bar represents the average from three replicates and standard error
        after the components have been oven dried except the seed   136

8.9   Seasonal crop water use versus (a) seed yield and (b) total revenues
        under land limiting conditions   138

8.10   Number of irrigations versus (a) production cost for the two
        scenarios and (b) economic water productivity under water limiting
        conditions   138

9.1   Daily rainfall at the experimental fields from April 2011 to February
        2014   143

9.2   Daily air temperatures at the experimental fields from April 2011 to
        February 2014   143

9.3   Daily relative humidity at the experimental fields from April 2011 to
        February 2014   144

9.4   Daily reference evapotranspiration at the experimental fields from
        April 2011 to February 2014   144

9.5   Simulated yield of the crop for full irrigation during the calibration
        of the model by using measured data of the 2013 irrigation season   149

9.6   Simulated dry biomass of the crop for full irrigation during the
        calibration of the model by using the 2013 irrigation season data   149

9.7   Relationship between simulated and measured canopy cover for full
        and deficit irrigation conditions during (a) calibration in the 2013
        and (b) validation in the 2013/2014 irrigation seasons. Each dot with
        standard error represents the mean of the measured canopy cover
        from two replicates   151

9.8   Relationship between simulated and measured dry above ground
        biomasses for full and deficit irrigation during (a) calibration in the
        2013 and (b) validation in the 2013/2014 irrigation seasons. Each dot
        with standard error represents the mean of the measured dry biomass
        from two replicates   152

9.9   Comparison of the simulated and measured soil moisture for full and
        deficit irrigation at depth of 0 - 0.8 m during the 2013 irrigation
        season. Each dot represents the mean of the soil moisture from two
        replicates. Standard error shows the deviation of the soil moisture
        from the mean value   153

9.10   Measured and simulated seed yields of Soybeans for the 2013 and
        2013/2014 irrigation seasons. Each dot represents the mean seed
        yield with standard error from three replicates   154

9.11   Measured and simulated dry above ground biomass of Soybeans for
        the 2013 and 2013/2014 irrigation seasons. Each dot represents mean
        of the dry above ground biomass with standard error from two
        replicates of the treatments   155

9.12   Measured and simulated water productivity of Soybeans for the 2013
        and 2013/2014 irrigation seasons   155

9.13   Measured and simulated cumulative crop evapotranspiration for
        Soybeans for (a) 2013 and (b) 2013/2014 irrigation seasons. Each
        dot represents the mean of the crop evapotranspiration with standard
        error from two replicates   156

D.1   Average soil temperatures in the upper 30 and lower 30 cm of the
        soil profile under different conservation practices and the
        conventional method in the rainy season   193

F.1   Partitioning of evapotranspiration into transpiration and evaporation
        for water conservation in 2012   211

List of Tables

Table	Description	Page
2.1	Agro-ecological zones in Nigeria (Food and Agriculture Organization of the United Nations (FAO), 2010c)	10
2.2	Average agricultural land area in Nigeria (1961-2000) (Oyekale, 2007; FAO, 2013)	11
4.1	Average area harvested, yield and production of Soybeans in Nigeria for the past 51 years	30
4.2	Some characteristics of Soybean breeding locations in Sub-Saharan Africa	31
4.3	The ranges of maturity, grain yields of late maturing promiscuous Soybean developed by the International Institute for Tropical Agriculture (IITA) in the Guinea savannah of West Africa	32
6.1	Summary of the duration (days) of the key phonologic stages of the crop in each year	79
7.1	Meteorological data measured at the experimental fields during the rainy seasons (standard deviations in parenthesis)	83
7.2	Physical and chemical properties of the soil at the experimental fields during the rainy seasons	85
7.3	Average plant heights at five different growth stages under rainfed conditions	90
7.4	Average number of leaves at five growth stages under rainfed conditions	91
7.5	Leaf area indices at different growth stages of the crop under rainfed conditions	93
7.6	Average fIPAR at different growth stages of Soybean during the rainy seasons	95
7.7	Comparison of RUEs determined by using PARs from global solar radiation and instantaneous measurements of PARs during the rainy seasons	96
7.8	Biomass at harvest, seasonal crop water use, water use efficiency for biomass production ($WP_{biomass}$) and transpiration efficiency (TE) in the rainy seasons	105
7.9	Yields and water productivity (seed) under rainfed conditions	112
7.10	Economic analysis of the production of Soybean under rainfed conditions	116
8.1	Meteorological data measured at the weather station located near the experimental fields in the dry seasons (Standard deviation in parenthesis)	123
8.2	Physical and chemical properties of the soil during the irrigation seasons	124
8.3	Leaf area index ($m^2\ m^{-2}$) at flowering, pod initiation, seed filling and maturity during the irrigation seasons	126
8.4	Dry matter ($g\ m^{-2}$) accumulation during the irrigation seasons	127
8.5	Number of seeds per plant and pods, maximum number of leaves in the mid season and maximum plant heights at harvest	128

8.6   Growth stages and their actual evapotranspiration (mm), seasonal evapotranspiration (mm) and number of irrigations in the 2013 and 2013/2014 irrigation seasons   129

8.7   Seasonal evaporation, transpiration, crop water use and seed yields in the two irrigation seasons   132

8.8   Water productivity, irrigation water productivity and harvest indices for full and deficit irrigation   135

8.9   Economic evaluation of the use of the drip method in cultivating Soybeans under full and deficit irrigation conditions in Ile-Ife   137

9.1   Conservative parameters used in simulating the response of Soybeans (Raes et al., 2012)   145

9.2   Non-conservative parameters used in simulating the response of Soybeans to water in Ile-Ife conditions   147

9.3   Results of the calibration of seed yield, dry above ground biomass under full and deficit irrigation conditions and percentage deviations of the simulated data from the measured values   154

Acknowledgements

I am thankful that I was born in the humble family of Pa Daniel Adegboyega Adeboye and Mrs. Olajiire M. Adeboye (late). Adage says that the journey of a thousand miles begins with a step. I appreciate my parents for giving me a good and lasting legacy in education by enrolling me in the prestigious public school in the late 1970s in Nigeria. I appreciate the teachers who taught me the rudiments of reading and writing at St Stephen's "B" Primary School Modakeke-Ife in Osun State in Nigeria under the leadership of Mr Ayodele the headmaster in 1985. My educational profile cannot be complete without appreciating the former Principal of Modakake High School, Modakeke-Ife under the headship of late Mr. S.O. Oloyede for his fatherly care and effort in laying a good foundation for his students especially those who graduated in 1991. I took a step of faith by travelling to the Netherlands and I am thankful my period of study in the Netherlands and the reward of my labour day and night.

I appreciate the useful suggestions of Prof. Dirk Raes of the Catholic University, Belgium and Prof. A. A. Olufayo of the Department of Agricultural and Engineering, Federal University of Technology, Akure, Nigeria during the writing of the proposal of this PhD thesis. I acknowledge the visitations of Prof. A.A. Olufayo, Dr. P. Oguntunde and Prof. M.O. Alatise of the Department of Agricultural Engineering, Federal University of Technology, Akure, Nigeria to my experimental fields and their pieces of advice, constructive criticism and useful suggestions for improvement.

Similarly, I acknowledge the effort, technical support and pieces of advice from Mr. A. Jejelola of the Institute of Agricultural Research and Training (IAR&T), Ibadan, Obafemi Awolowo University (OAU) House Station during the entire fieldwork at the Teaching and Research Farms of OAU, Ile-Ife. In addition, I appreciate the technical services rendered by the staff members of the Farm Office at Teaching and Research Farms of OAU, Ife during this study. I thank Mr. A.A Adeleke and Mr. Udor of the Crop Research and Seed Processing Unit of the International Institute for Tropical Agriculture (IITA) Ibadan for providing advisory services on the cultivar of Soybean used in this study. I appreciate Prof. O.O. Jegede and Mr. M. Bashiru of the Department of Physics, Obafemi Awolowo University for providing Meteorological data that were used in complementing the data measured in the field during this PhD study.

I appreciate the effort of the final year students in the Department of Agricultural and Environmental Engineering, Obafemi Awolowo University who laboured with me during the fieldwork at the Teaching and Research Farms. The students are: Chimezie, Joy Oluwaseun; Mayowa Oladimeji, Adeyinka Funmilayo Mofoluwasola; Jegede Yussuf; Falana Olumide and Bolatito Abiola. I wish you the best in your live endeavours.

I acknowledge the Dutch government through the Department of Development Cooperation and Netherlands Fellowship Programme (NFP) for financing my Doctoral Degree programme at UNESCO-IHE Institute for Water Education, Delft, the Netherlands. I appreciate the support of the management of UNESCO-IHE and friendliness of the staff members of the institute. I encourage you to keep it up. I appreciate the management of Wageningen University and Research Centre, for their academic support.

I appreciate my Promotor, Em. Prof. Dr Ir. E. Schultz for his full support and appropriate guidance on all practical matters throughout my study at UNESCO-IHE. I appreciate his perseverance, accommodation and the excellent academic trainings received from him throughout the period of my study in the Netherlands. I will forever remember his contributions to my academic career and capacity development. Similarly I appreciate my mentor Dr. Krishna Prasad from Nepal for all his efforts now in and building my write up from proposal to the thesis stage. Working with him was a wonderful and rewarding experience for me. I appreciate his kindness and ingenuity. I wish you the best in your entire endeavour in life. I acknowledge the effort of my local supervisor, Prof. K.A. Adekalu for his advice and academic support throughout the period of this study.

I thank my dear wife Mrs. Adeboye Amaka Precious for her support and assistance during this study. I acknowledge the moral supports from members of my family Pa D.A. Adeboye, Mr. A.A. Adeboye, Mr. F.I. Adeboye during the study.

I appreciate the financial contributions and technical supports from Robert B. Daugherty Water For Food Institute of the University of Nebraska, Lincoln, United States of America (USA), European Union (EU) and the Food and Agriculture Organization of the United Nations (FAO) during my PhD study. I appreciate the management of Obafemi Awolowo University, Ile-Ife for their support during my period of study in Europe.

Summary

Human population is increasing faster than ever in the history. There is an urgent need to scale up food production in order to meet up with food demands, especially in Sub-Saharan Africa. In Ogun-Osun River Basin, Nigeria, more than 95% of the crop production is done under rainfed conditions. Fluctuation in rainfall as a result of climate change is a major challenge in the recent times in the basin. Land productivity can be greatly improved by using affordable water conservation practices by peasant farmers who produce crops in the basin. Similarly, water saving measures would have to be adopted by using drip irrigation and application of water at critical stages of growth of crops. Fertility of the soil needs to be maintained by cultivating crops that naturally replenish soil nutrients. Such measures will go a long way in ensuring sustainable use of land and water in Ogun-Osun River Basin.

An indeterminate cultivar of Soybeans TGX 1448 2^E was cultivated at the Teaching and Research Farms of Obafemi Awolowo University, Ile-Ife, Nigeria during the rainy seasons from May to September, 2011 and June to October, 2012. Similarly, the crop was drip irrigated for two dry seasons from February to May in 2013 and from November, 2013 to February, 2014. The purpose of conducting the experiments in the rainy and dry seasons was to compare the yields and their components and to evaluate the performances of the crop in terms of water use and productivity. The experimental field during the dry season was located at about 1 km from the field used during the rainy season due to the nearness to the source of water. During the experiments in the four seasons, key biometric data of the crop were taken from emergence to physiological maturity. The crop cycle during the rainfed experiment lasted for 117 and 119 days in 2011 and 2012 respectively, while in the dry season it lasted for 112 days in the first season and 105 days in the second season. The lengths of the crop cycles in the four seasons differed a little bit. This is attributed to environmental factors such as weather conditions, nutrient availability in the soil and period of cultivation. During the rainy seasons, six water conservation treatments were used namely Tied ridge, Mulch, Soil bund, Tied ridge plus Soil bund, Tied ridge plus Mulch, Mulch plus Soil bund and Direct sowing without water conservation measure (conventional practice), which was the control treatment. The treatments were placed in a randomised complete block design with four replicates in an area of 31 by 52 m (1,612 m²) and standard agronomic measures were taken. Soil water balance approach was used in determining evapotranspiration during the rainfed and irrigation seasons. Seasonal evapotranspiration was partitioned into the productive transpiration from the plants and non-productive evaporation from the soil.

Seasonal average canopy extinction coefficients were 0.46 and 0.51 respectively in the rainy seasons of 2011 and 2012, while in the dry seasons of 2013 and 2013/2014 they were 0.43 and 0.49. The plant height ranged from 51.3 cm for Soil bund to 67.8 cm for the conventional practice in 2011 while in 2012, it ranged from 60.3 cm for Tied ridge plus Soil bund to 80.3 cm for Mulch plus Soil bund. The minimum fraction of Intercepted Photosynthetically Active Radiation was 0.13 during establishment for Tied ridge plus Soil bund while the peak fraction was 0.97 during seed filling for Soil bund during the rainy seasons. Similarly, the minimum and peak leaf area indices were 0.13 $m^2 \, m^{-2}$ for Tied ridge plus Soil bund during establishment in 2011 and 6.61 $m^2 \, m^{-2}$ for Soil bund during seed filling in 2012. There were strong and significant correlations between the fraction of Intercepted Photosynthetically Active Radiation and the leaf area indices (LAI) $(0.70 \le r^2 \le 0.99)$ in 2011 and $(0.93 \le r^2 \ge 0.99)$ in 2012 by using an exponential model. Seasonal rainfall in 2011 and 2012 was 539 and 761 mm respectively. Seasonal water storages in the soil in 2011 ranged from 407 mm for the conventional practice to 476 mm for Tied ridge plus Mulch, while in 2012 it ranged from 543 mm for Tied ridge to 578 mm for Tied ridge plus Soil bund.

Radiation Use efficiency was determined by plotting dry above ground biomass measured at intervals of seven days against the Daily Photosynthetically Active Radiation from Solar radiation and the Instantaneous Photosynthetically Active Radiation measured near solar noon for all the treatments. For the Photosynthetically Active Radiation obtained from solar radiation, Radiation Use Efficiency of the crop ranged from 1.18 g MJ^{-1} for Tied ridge to 1.98 g MJ^{-1} of Intercepted Photosynthetically Active Radiation for Tied ridge plus Soil bund in 2011, while in 2012 it ranged from 1.45 g MJ^{-1} for Tied ridge to 1.92 g MJ^{-1} for Mulch. There was no significant difference in the average seasonal Radiation Use Efficiency in the two seasons. By using instantaneous measurement of the Photosynthetically Active Radiation, Radiation Use Efficiency ranged from 0.80 g MJ^{-1} of Intercepted Photosynthetically Active Radiation for Tied ridge to 1.65 g MJ^{-1} for Tied ridge plus Soil bund in 2011, while in 2012 it ranged from 0.94 g MJ^{-1} for Tied ridge to 1.24 g MJ^{-1} for Soil bund. The two approaches gave relatively similar values of Radiation Use Efficiency. Positive - correlation coefficients ($0.50 \leq r^2 \leq 0.89$) were found among the treatments between the dry above ground biomass simulated by using a light model and those measured in the field in the two seasons.

The seasonal crop water use ranged from 311 mm for Mulch plus Soil bund to 406 mm for Tied ridge plus Soil bund in 2011, while in 2012 it ranged from 533 mm for Mulch plots to 589 mm for Soil bund. Seasonal transpiration ranged from 190 mm for Tied ridge plus Mulch to 204 mm for Soil bund in 2011 while in 2012 it ranged from 164 mm for Tied ridge plus Mulch to 195 mm for Mulch plot. Seasonal evaporation was higher in 2012 ranging from 338 mm for Mulch plots to 408 mm for Soil bund while in 2011 it ranged from 311 mm for Mulch plus Soil bund to 406 mm for Tied ridge plus Soil bund. Water storage in the soil and seasonal crop water use are significantly related. Similarly, the seasonal crop water use, Intercepted Photosynthetically Active Radiation and Radiation Use efficiency were highly related for the crop over the two seasons.

Marketable seed yield ranged from 1.68±0.50 t ha^{-1} for Tied ridge to 2.95±0.30 t ha^{-1} for Tied ridge plus Soil bund in 2011, while in 2012 the yield ranged from 1.64±0.50 t ha^{-1} for the conventional practice to 3.25±0.52 t ha^{-1} for Mulch plus Soil bund. In 2011, seed yield for Tied ridge plus Soil bund was 15.6, 15.9, 25.4, 28.5, 43.1 and 47.1% higher than seed yield for Mulch plus Soil bund, Soil bund, Mulch, Tied ridge plus Mulch, Tied ridge and conventional practice respectively. In 2012, seed yield for Mulch plus Soil bund was 7.4, 21.8, 32.0, 32.3, 43.7 and 49.5% higher than the seed yields for Soil bund, Tied ridge, Mulch, Tied ridge plus Mulch, Tied ridge plus Soil bunds and Direct sowing respectively. Average seasonal seed yield of the crop was significantly related to the Total Intercepted Photosynthetically Active Radiation but not to the Radiation Use Efficiency. Harvest indices ranged from 47.4±4.5% for Tied ridge to 57.6±1.1% for Tied ridge plus Soil bund in 2011 and 53.1±3.0% for Soil bund to 58.1±2.3% for Tied ridge 2012. The highest harvest indices were obtained in Tied ridge plus Soil bund and Tied ridge in 2011 and 2012 respectively. Harvest index was not significantly related to both Intercepted Photosynthetically Active Radiation and Radiation Use Efficiency of the crop.

Average seasonal transpiration efficiencies - the ratio of the dry above ground biomass at harvest to the seasonal transpiration - for all the treatments were 7.0 kg ha^{-1} mm^{-1} in 2011 and 14.9 kg ha^{-1} mm^{-1} in 2012. Transpiration efficiency of the crop was strongly related to Intercepted Photosynthetically Active Radiation but not to Radiation Use Efficiency under field conditions in the rainy seasons. The peak water productivity for seed was 7.99 kg^{-1} ha^{-1} mm^{-1} in 2011 and 5.76 kg^{-1} ha^{-1} mm^{-1} for Mulch plus Soil bund in 2012. Water productivity for seed was strongly and significantly related to Intercepted Photosynthetically Active Radiation. However, it was not significantly related to Radiation Use Efficiency. These findings will provide information to the crop yield modellers during the simulation of yields of Soybeans under water conservation practices.

The construction of ridges and Soil bund especially for Tied ridge, Mulch plus Soil bund and Tied ridge plus Soil bund increased the average seasonal cost of production by 28.9% compared with Mulch and conventional practice and by 10.1% compared with Soil bund. In addition, economic water productivity was 3.90 US$ ha^{-1} mm^{-1} for Mulch plus Soil bund while for Soil bund and conventional practice, it was 3.30 and 2.27 US$ ha^{-1} mm^{-1} respectively.

Due to increase in demand for food, there is the need to produce more crop per drop of water under rainfed conditions and to manage water for agriculture at basin scale. The key priority in the study area was to increase the seed yields, water and economic productivity and the financial benefits at the end of a cropping season. The results show that the use of Mulch plus Soil bund had the average maximum transpiration efficiency, seed yield, water and economic productivity, and revenue of 1,630 US$ per ha. By comparing the average seasonal transpiration efficiency, crop water use, yield, water productivity and costs of production for the six conservation practices with those of the conventional practice in the two rainy seasons, Mulch plus Soil bund had the maximum average seed yield, water and economic productivity. Mulch plus Soil bund is hereby recommended for the cultivation of the crop in the study area. Other conservation practices, such as Soil bund, also performed satisfactorily in terms of seed yield and water productivity, although with a slight reduction in revenue. The use of these water conservation practices will not only increase the yields of the crop, but reduce depletion of water in the soil, which could initiate or increase land degradation in the study area to the barest minimum. Hence, sustainability of land and water in Ogun-Osun River Basin can be ensured. These findings demonstrate that land and water productivity of Soybean under rainfed conditions can be significantly improved with water conservation practices under the current fluctuations of rainfall and competition for land resources between agriculture and urban land use in Ogun-Osun River Basin.

Field trials were also conducted for two irrigation seasons from February to May, 2013 and November, 2013 to February, 2014. The crop was planted in a Randomized Complete Block Design with three replicates and in-line drip irrigation was applied to supply water to the crops. Five treatments were selected and these are: (i) full irrigation, skipping of irrigation every other week during (ii) flowering; (iii) pod initiation; (iv) seed filling and (v) commencement of maturity. Biometric data, which are number of leaves, plant height, leaf area indices and dry above ground biomass, were taken and recorded every week from sowing until maturity in the two irrigation seasons. Soil moisture contents were taken at the root zone of the plants prior to irrigation in order to determine the net irrigation water requirements at each stage of growth. Harvest indices were determined for each treatment. Number of pods per plant, number of seeds per pod and yields under each treatment were determined after physiological maturity in each season. Regression equations were generated for: (i) yield; (ii) number of pods per plant; (iii) number of seeds per pod; (iv) number of leaves; (v) seasonal transpiration and leaf area indices. Similarly, regression equations were generated for: (i) plant heights; (ii) seasonal transpiration; (iii) number of pods per plant; (iv) number of seeds per pod; (v) dry above ground biomass. Linear regressions were also fitted to the yield, dry above ground biomass and seasonal crop water use. The crop response factor was determined. Water productivity and Irrigation water productivity were computed and compared for each treatment. Linear models were fitted to the water productivity, irrigation water productivity and harvest index.

Rainfall contribution to the crop water use was 262 and 50 mm for 2013 and 2013/2014 irrigation seasons respectively. Maximum Leaf Area Index in the 2013 irrigation season was 7.10 m^2 m^{-2} for full irrigation during seed filling, while in the 2013/2014 irrigation season, it was 3.44 m^2 m^{-2} for full irrigation during flowering. The dry above ground biomass after maturity ranged from 359 g m^{-2} where irrigation was skipped every other week at the commencement of maturity to 578 g m^{-2} for full irrigation. The seed yields ranged from 1.81 t ha^{-1} when irrigation was skipped every other week during seed filling to 3.11 t ha^{-1} for full irrigation. Average seasonal seed yield for full irrigation was 18.8, 21.8, 24.4 and 47.9% higher than yields for treatments where irrigation was skipped every other week during flowering, pod initiation, commencement of maturity and seed filling respectively. Seasonal transpiration ranged from 217 mm when irrigation was skipped every other week during seed filling to 409 mm for full irrigation in the 2013 irrigation season, while in the 2013/2014 irrigation season it ranged from 28 mm for the treatment where irrigation was skipped every other week during seed filling to 223 mm for full irrigation. Seasonal crop water use ranged from 463 mm when irrigation was skipped every other week during flowering to 523 mm for full irrigation in the 2013 irrigation season, while in the 2013/2014 irrigation season it ranged from 364 mm when irrigation was skipped every other week during seed filling to 507 mm for full irrigation. Harvest indices ranged from 56.0% when irrigation was skipped during seed filling to 65.9% when irrigation was skipped during flowering in the 2013 irrigation season, while in the 2013/2014 irrigation season, it ranged from 43.2% when irrigation was skipped during seed filling to 63.9% for full irrigation. Water productivity for seed production ranged from 3.89 kg ha mm^{-1} when irrigation was skipped during seed filling to 5.95 kg ha^{-1} mm^{-1} for full irrigation in the 2013 irrigation season while in the 2013/2014 irrigation season, it ranged from 1.93 kg ha mm^{-1} when irrigation was skipped during seed filling to 3.00 kg ha^{-1} mm^{-1} for full irrigation. Irrigation water productivity ranged from 8.90 kg ha mm^{-1} when irrigation was skipped during seed filling to 14.0 kg ha^{-1} mm^{-1} when irrigation was skipped during flowering in 2013, while in the 2013/2014 irrigation season, it ranged from 2.24 kg ha^{-1} mm^{-1} when irrigation was skipped during seed filling to 3.32 kg ha^{-1} mm^{-1} for full irrigation. Leaf area indices and yield, number of leaves, number of pods per plant, number of seeds per pod and seasonal transpiration were significantly correlated. Similarly, dry above ground biomass and seasonal transpiration, number of pods per plant, number of seeds per pod were significantly correlated. The crop response factor (K_y), a measure of the relative decrease in seed yield due to relative decrease in evapotranspiration, was 2.24. It indicates that the deficit irrigation imposed on the crop was high and that relative decrease in yields due to deficit irrigation was higher than relative decrease in evapotranspiration.

Results show that skipping of irrigation at any growth stage of the crop led to reduction in the leaf area indices, dry above ground biomass and seasonal crop water use. Deficit irrigation had significant effects on both the dry matter and yields. The effect of deficit irrigation was more pronounced on seed yields than on dry matter. Severity of the effects of deficit irrigation depended on the stage of growth and its duration. Deficit irrigation reduced significantly dry matter at flowering and pod initiation. However, deficit irrigation did not affect the plant height. Number of seeds per plant at flowering and commencement of maturity were reduced significantly by deficit irrigation. The number of seeds per pod was significantly reduced when irrigation was skipped at pod initiation only. Seed yields were significantly reduced when irrigation was skipped during seed filling. In the 2013 irrigation season water productivity when irrigation was skipped during flowering was 2.3, 16.1, 23.5, and 36.1% higher than water productivity for full irrigation, when irrigation was skipped during pod initiation, commencement of maturity and seed filling respectively. In the same season, irrigation water productivity when irrigation was skipped during flowering was 15, 20, 29.3 and 36.4% higher than for full irrigation, when irrigation was skipped during pod initiation, commencement of maturity and seed filling respectively. In the 2013/2014 irrigation season, however, water productivity for full irrigation was 8.7, 16.3, 24.7 and 35.7% higher than when irrigation was skipped during pod initiation, commencement of maturity, flowering and seed filling respectively. Similarly, irrigation water productivity was 7.2, 15.4, 24.1 and 32.5% higher than when irrigation was skipped during pod initiation, commencement of maturity, flowering and seed filling respectively. In addition, irrigation water productivity for full irrigation was 24.1 and 32.5% higher than when irrigation was skipped during flowering and seed filling respectively. Stage of growth, its duration, water requirements and seasonal environmental conditions influenced the seasonal water use, water productivity and irrigation water productivity of Soybean. Maximum water productivity and irrigation water productivity were obtained when irrigation was skipped every other week during flowering only in the first season, whereas in the second season full irrigation gave the peak water and irrigation water productivity. This suggests that irrigation water productivity of Soybean can be improved upon by skipping irrigation during flowering and pod initiation.

    In this study, the costs of production for all the irrigation scenarios were high. This is due to the high cost of water, which constituted between 54 to 59% of the production cost if water is purchased and cost of drip irrigation equipment, which constituted between 75.6 to 76.7% of the total cost of production if water would be given without financial implication. Under the prevailing price and economic conditions after harvest, the use of in-line drip irrigation does not offer economic benefit to peasant farmers, who are the predominant growers of the crop in the study area. Economic benefit may be achieved after long periods of usage with proper maintenance of the irrigation facilities and elimination of the fixed cost from the total cost of production.

    The water driven crop model AquaCrop was calibrated and validated to predict canopy cover, dry above ground biomass, seed yield, evapotranspiration, soil moisture content and water productivity of the crop. The simulated and measured data compare adequately except for water productivity that was over predicted in the validation data set. The AquaCrop model predicted canopy cover with error statistics of $0.93 \leq E \leq 0.98$ for both full and deficit irrigation and the degree of agreement $d = 0.99$ with $4.3 \leq RMSE \leq 5.9$ (root mean square error) for full irrigation while for deficit irrigation, $0.96 \leq d \leq 0.99$ with $5.3 \leq RMSE \leq 5.8$. Dry above ground biomass was predicted with error statistics of $0.08 \leq RMSE \leq 0.14$ t ha^{-1} with $0.98 \leq d \leq 0.99$ for full irrigation, while for deficit irrigation it was $0.06 \leq RMSE \leq 1.09$ t ha^{-1} with $0.85 \leq d \leq 0.99$. One in every five predictions of the above ground biomass was outside 20% deviation from the measured values.

    The seed yields were predicted with error statistics of $RMSE = 0.10$ t ha^{-1} and $d = 0.99$ and one in five predictions was outside 15% deviation from the measured data. The prediction error statistics for seasonal crop water use for both full and deficit irrigation treatments was $15.4 \leq RMSE \leq 58.3$ in the two seasons. The AquaCrop model over predicted percolation also in the validation data set. These observations suggest that the percolation components of the model need to be adjusted to ensure better performance. The performance of the AquaCrop model in predicting canopy cover, seed yield and other quantities in this study are commendable and satisfactory.

Specific and distinct features, such as the use of canopy cover rather than leaf area index, make the model suitable for developing countries like Nigeria, where researchers may not have access to state-of-the-art equipment for measuring the leaf area index. Similarly, water productivity that is normalized for atmospheric demand and carbon dioxide concentration and its focus on water makes it suitable for diverse locations. Over the years, it has been observed that no model is universal in its ability to take into consideration all differences in cultivar, environment, weather and management conditions. Other cultivars of Soybeans in Nigeria and other agro-climatic environments need to be tested and fine-tuned in the model, in order to ascertain the accuracy of the model. Generally, the model predicted the stated parameters with reasonable degree of accuracy and is hereby recommended for use in Ile-Ife and other parts of Ogun-Osun River Basin and Nigeria.

Although land, water, and economic productivity of the crop were higher where water was conserved under rainfed conditions, treatment of the soil to conserve water and regular maintenance increased the average seasonal cost of production compared with the conventional practice. High cost of production may reduce the benefits obtained by the crop growers, except when there is improvement in the market price. Therefore, sustainable practice of the water conservation measures must be accompanied with lower cost of production. Under irrigation conditions, the land and water productivity are lower compared with rainfed cultivation. The productivity in the dry season reduces with the severity of the water stress. Average crop water productivity and economic water productivity of all the six water conservation measures in the rainy season were higher than with full irrigation in the dry season. The costs of production of the crop in the dry season were significantly above the cost during the rainfed conditions. Higher water productivity under rainfed conditions in this study is in agreement with the finding that in a significant part of the least developed and emerging countries there is larger opportunity for improving water productivity under rainfed conditions compared to irrigated agriculture.

Expansion of arable land may not be feasible in Ile-Ife because of the huge investments involved. Thus, the focus of efforts to expand food production in the area would have to be on raising land productivity on the existing arable lands and improving production efficiencies, outcomes that can only be achieved by using improved cultivars together with improved agronomic practices. Agronomic practices, especially under rainfed conditions, would have to be designed to improve water productivity. Improving water productivity requires vapour shift (transfer) whereby soil physical conditions, soil fertility, crop varieties and agronomy are applied in tandem and managed to shift the evaporation into useful transpiration by plants. During the dry season, the crop would have to be irrigated in order to achieve maximum land and water productivity. Skipping of irrigation during seed filling would have to be avoided in order to prevent significant reduction in yield. Irrigation at the commencement of maturity after the pods have been completely filled with seeds can be skipped. Under water limiting conditions, the amount of water saved by skipping irrigation during flowering, pod initiation, seed filling and maturity can be used for cultivating other crops and thereby increasing the opportunity cost. Incidental rainfall during the dry season would have to be used in order to increase irrigation water productivity of the crop.

1 Introduction

Water is the most abundant natural resource on the earth and a major substance that ensures continuity of ecosystems and biodiversity. Since millennia, man has been by using water to his advantages in diverse ways and at the same time has protected himself against the harmful effects of water purposely to improve his living conditions. On regular basis, life depends on water and its path of flow determines the shape of the earth. Sustainable welfare of man and all living beings depends on wise and safe use of water resources. Historically, land and water have notable contributions to social and economic development of all regions of the world. Irrigated agriculture along River Nile in Egypt and hydroelectric power generation at Kainji Dam in Nigeria are few of numerous examples (Ray et al., 1988). In the West African Sub-region, the Senegal and Niger rivers play prominent roles in enhancing agricultural activities. In Europe for instance, the Rhine valley, which is recognized as a locus of both co-operation and conflict was a primary nexus of economic growth (Sadoff and Wittington, 2002).

Water finds application in diverse ways. These include domestic, agricultural, industrial, recreation and nature conservation uses. Apart from its industrial uses, water is a very essential social amenity. The provision of clean portable water in Nigeria has gone a long way in reducing water-borne diseases and in improving the general sanitation of towns and cities. Despite the relative abundance of water, the complaints everywhere are the same, 'shortage of supply in quantity and quality'. In many of the African countries and in other places, the demand for water has been on the increase (Sharma et al., 1996). This can be attributed to the increase in human population and extensive migration caused by economic pressure and natural disasters. The demand for water and productive land will be on the increase as well in the nearest future (Neil, 1995). Due to water scarcity, poverty and stressed ecosystems, about 850 million people live in conditions associated with food insecurity (Food and Agriculture Organization of the United Nations (FAO), 2009). In Nigeria, only 60% of the population has access to improved drinking water out of which 49% of the rural population has access to safe water. Similarly, an expected additional 1-2 billion people will need to be fed by 2025 (United Nations (UN), 2009). This places a demanding challenge on water resources especially in areas where water is scarce or water resources are not exploited fully because a large amount of water is required for food production. According to Caroline (2002) global water resources are limited and only through a more sustainable approach to water management more equitable and ecological sensitive strategies of water allocation and use, can we hope to achieve the international development targets for poverty reduction that have been set for 2015.

The world contains an estimated 1,400 million km^3 of water but only 45,000 km^3 (0.03%) is regarded as fresh water, that is the water that can be used for drinking, hygiene, agriculture and industry, while the remaining proportion is saline water. About 9,000 to 14,000 km^3 are economically available for human use, thereby making fresh water a very valuable and scarce resource (Food and Agriculture Organization of the United Nations (FAO), 2010a). 75% of the earth's fresh water is contained in ice caps and glaciers, while another 14% is locked up in very deep and inaccessible aquifers as reported by the Commission on Sustainable Development (CSD) (2002). From the total volume of the available fresh water resources, about 20% is used by industries, 10 by household and 70% are used by agro allied industries respectively (Commission on Sustainable Development (CSD), 2002; Cai and Rosegrant, 2003). The breakdown of the fresh water use in the whole world is shown in Figure 1.1. In Africa, 84.1% of water is consumed by agriculture while 8.6 and 7.3% are used for domestic and industrial purposes. This is large when compared with only 32.4% being consumed by agriculture in Europe.

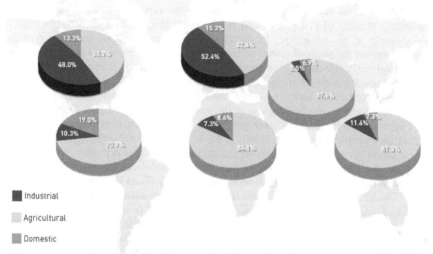

Figure 1.1. Global freshwater use (Food and Agriculture Organization of the United Nations (FAO), 2010b)

Ogun-Osun River Basin is located in the sub-humid tropical area of Nigeria. Cultivation of crops in the basin is done mainly in the rainy season at subsistence level. Varieties of crops such as Maize, Soybeans, Plantain, Banana, and Yam are cultivated during the rainy season. In the dry season, crops that are cultivated include Maize and vegetables such as Amaranthus. The rainy season spans between March and October while the dry season spans between November and February. There are fluctuations in the recent times that can be attributed to effects of changes in climate in the basin. The variability in rainfall dictates the period of farming. The crop cultivation in the dry season is done at lowland or waterlogged areas under the current Fadama Program. Crop cultivation is done by illiterate or semi-literate farmers whose productivity is very low due to low farming input. Production of these crops is reducing due to the fewer number of people that are involved in crop production.

The increase in human population and effects of climate change are mounting heavy pressure on freshwater resources in the basin and therefore there is a need to devise means of copping with the effects of fluctuations of rainfall on crop production. This can be done by introducing innovative and affordable practices of water conservation measures in rainfed farming. This study examines the effects of six water conservation and the conventional practices on yield and yield components of Soybeans *(Glyxine max. L. Merr.)* at the Teaching and Research Farms of Obafemi Awolowo University, Ile-Ife, Nigeria. The conservation practices are the use of Mulch (plant materials), Tied ridge, Soil (side) bund, Direct sowing without any conservation measure, Soil bund plus Tied ridge and Mulch plus Soil bund. Field trials were conducted in the rainy seasons of 2011 and 2012 and yields and yield components obtained have been compared. The crop water use of Soybean was determined and impacts of the water conservation measures on improving land and water productivity (WP) were examined.

Similarly, Soybean was drip irrigated in the dry seasons of 2013 and 2013/2014 at the Teaching and Research Farms of Obafemi Awolowo University, Ile-Ife. Application of water was skipped during four sensitive reproductive stages of the crop, which are flowering, pod initiation, seed filling, and maturity. The purpose of skipping the water application at these stages was to save water, determine the stage(s) of the growth in which water stress will reduce yield of the crop and produce more crops per drop of water. Water use, WP and Irrigation water productivity (IWP) were determined and comparisons were made among the treatments. Growth, yields, and water productivity of the crop under the Ile-Ife conditions were simulated by using the AquaCrop model.

1.1   Structure of the thesis

In order to address the questions raised and achieve the desired objectives, this thesis is structured into ten chapters. The paragraphs below give an outline of the structure.

Chapter 1 gives a general introduction of the state of water and food production on global and local scales. It contains the scope and structure of the thesis.

Chapter 2 gives a general overview of the research. This includes concise explanations of its relevance, the questions that are addressed, scope and preliminary guidelines, the hypothesis, research objectives and methodology. It gives a general profile of Nigeria in terms of weather and climate, administrative divisions, their geographic locations and the state of land and water resources development. The established river basin development authorities are outlined and their effort, success and challenges in managing water resources are stated.

Chapter 3 presents a full description of the study area Ogun-Osun River Basin in terms of climate, rainfall pattern, land use practices, soil and land resources. This chapter ends with a comprehensive overview of the current problems and challenges envisaged in the basin in the nearest and distant future and the relevance of this research in addressing those challenges.

Chapter 4 gives detailed information on the Soybean under investigation. A brief history of the origin of the plant and its distribution is stated in this chapter. The biological description of the plant is stated in order to aid the understanding of its agronomy, which will assist and serve as a guide to the researcher in monitoring the crop while on the field. The method of propagation and management of the plant on the field is also included in this chapter. The diseases and pest identified with this crop and an account of the recorded yield in Nigeria, Zimbabwe and other places in the world is stated. This chapter ends by a review of the time of planting and the available cultivars in Nigeria based on recent researches at the International Institute of Tropical Agriculture (IITA), Ibadan.

Chapter 5 explores the literature that is relevant to various crop yield models. It gives a comprehensive review of the categories of models that are available in literature. It explains their application and limitations. Description of the AquaCrop model is also included.

Chapter 6 explains in detail the research methodologies used on the experimental fields during the rainy and dry seasons. It explains in detail the experimental treatments, cultivation, field management approaches and the measurements made in the fields. The chapter also contains the daily observations and records of growth and phenological development of the plant throughout the fieldwork in the rainy and dry seasons.

Chapter 7 contains in detail the results and discussions on the data obtained during the rainfed experiments in 2011 and 2012. The analysed biometric data measured in the field and the implications of the results include the leaf extinction coefficient, leaf area index, dry matter, harvest index, yield and crop water use, as well as the effects of water conservation on soil water storage and yield of the crop.

Chapter 8 contains results obtained during the dry seasons of 2013 and 2013/2014. The effects of deficit irrigation on biometric data and the relationship between accumulation of dry matter, yield and irrigation water applied are stated and explained. Data on seasonal evaporation, transpiration and crop water use under different treatments are compared in order to identify the treatment with the highest WP and the effects of skipping irrigation at a particular stage of the growth. Crop response factors were determined for specified stages of growth and their implications are explained.

Chapter 9 contains procedures and results obtained by using AquaCrop to model the yield and growth of the crop in response to water stress under deficit irrigation conditions. In addition, methods of evaluating the performance of the model are stated in this chapter. Furthermore, this chapter contains considerations of anticipated challenges when implementing the optimal solution and recommendations stated in chapters 8 and 9 at the local and regional levels, especially among the peasant and rural farmers in Ogun-Osun River Basin.

Chapter 10 contains the evaluation and conclusions on the knowledge gained from the study.

1.2   Scope of the thesis

In order to limit the research area to an in-depth manner, more relevant, easier to comprehend, and the findings that are easier to apply, the scope has to be defined and limited as follows. Ogun-Osun River Basin covers five states namely: Lagos, Oyo, Ogun, Osun and part of the present Kwara states in Nigeria. Available background information and data in this study cover the entire basin. The current and future challenges in this study also reflect the conditions in the entire basin.

Conducting the research in the entire basin was not required. Therefore, the Teaching and Research Farms of Obafemi Awolowo University, Ile-Ife was selected as experimental site. Ile-Ife is located at the centre of the basin. The site was selected because of the nearness to research facilities such as arable land, and a dam for irrigated agriculture. Many annual and perennial crops are cultivated in the basin, but only Soybean (a cultivar) highly embraced by the Soybean growers and IITA was cultivated during the period of the study. Soybean was selected because of its importance in the daily diet of the people in the basin and the decision of the Ministry of Agriculture to encourage the large-scale production of the crop.

The study was conducted during two rainy and two dry seasons and a comparison was made between the yields, land and WP of the crop at the experimental site. The experimental treatments in the rainy seasons focused on the use of water conservation practices in cultivating Soybeans but not on the comparison of the productivity of different cultivars. The water conservation practices were selected because of the environmental challenges such as fluctuations in rainfall due to the impacts of climate change in the basin. Measurements made at the experimental fields were biometric (leaf area index, canopy cover, plant height, number of leaves, leaf extinction coefficient, seed and biomass yields), soil analysis (physical and chemical properties) and meteorological data. A detailed study of the soil, effects of climate change on the crop in the basin and analysis of water quality used at the experimental site was outside the scope of this study.

In the dry season, the crop was drip irrigated in the same experimental area. Water deficit was limited to sensitive reproductive growth stages only namely:

flowering, pod initiation, seed filling and maturity in order to examine the effects of water stress on the crop. Seasonal crop water use, evaporation and transpiration, harvest indices, water productivity, transpiration efficiency, crop response factor, water use and irrigation water use efficiencies were determined.

AquaCrop, a dynamic and water driven productivity model was used in simulating the effects of full and deficit irrigation practices on canopy growth, accumulation of biomass, soil moisture, yield and water productivity of the crop.

Symbols are shown in Appendix A and Acronyms in Appendix B. The Dutch summary is shown in Appendix Q and finally information about the author is given in Appendix R.

2 Background and objectives

2.1 Current challenges in water resources development and management

Appropriate actions, according to recent forecasts need to be taken to improve water management and to increase water use efficiency (Alcamo et al., 1997; Seckler, 1996; Shiklomanov, 2000; Rosegrant et al., 2002, 2005; Bruinsma, 2003; Falkenmark and Rockstrom, 2003; Vorosmarty et al., 2004). Pollution and land degradation are other environmental factors that reduce the availability of clean fresh water in almost all parts of the world. This is manifest during mining activities and poor management of agricultural land. In Africa, land degradation and extensive desertification are direct effects of intensive land cultivation and inappropriate land use systems (Mohammed et al., 1996). In addition, changes in global climate are imminent challenges facing crop production and the environment. Various social and economic activities upset the natural hydrologic balance in the least developed countries where the natural resources have not been fully developed and utilized. Deforestation and lumbering activities expose land surfaces to the battering action of tropical rainfall thereby initiating soil erosion and sediment accumulation. This is one of the major causes of frequent flood disasters in Nigerian coastal cities such as Lagos and Port Harcourt. Poor urban planning and dumping of agricultural and industrial wastes in water channels are other causes of river flooding (Morgan, 1996). Uncontrolled application of synthetic fertilizers and manure from livestock production sludge in municipal sewage treatment plants and waste from agricultural activities can have many negative impacts on water quality in any river basin. Manure application from livestock and direct runoff may lead to soil acidification, which in turn may increase metal solubility in soils (Food and Agriculture Organization of the United Nations (FAO), 1999). Removal of riparian vegetation causes increase in temperature, which reduces oxygen solubility and adversely affects biological activities in the water as well as self-cleaning capacity of a river. Likewise, soil compaction and increase in drainage capacity can also lead to an increase in peak flow. Other drivers are urbanization and human migration.

Diminished water allocation to agriculture is no longer a future challenge. Privatization of the groundwater market in China has led to increasing water scarcity (Zhang et al., 2008). In Guadalquivir River Basin in Spain, the water authority recently allocated less water to the irrigation district while an increase of 15-20% in irrigation water needs have been predicted for 2050 (Rodrignez-Diaz et al., 2007). Dependency on rainfall for future crop production has become a major constraint for sustainable food production in the emerging countries including Nigeria and China (Karam et al., 2007). With an increasing human population and less water availability for food production, food security for the future generation is at stake (Zwart and Bastiaanseen, 2004). Sustainable food and fibre production, which are expected to cater for the teeming population will depend largely on judicious and conjunctive use of surface and groundwater in order to attain the Millennium Development Goals (MDG) of equitable water distribution and usage for all by 2015 (Smith, 2000; Howell, 2001; Molden, 2003; United Nations World Water Assessment Programme (UNWWAP), 2006).

Agriculture is the largest water user. Irrigated agriculture accounts for the usage of about 90% of the available water resources in the least developed countries and 70% of the total water withdrawal worldwide. However, about 40% of the agricultural output is generated from irrigated agriculture despite very large acreage of land allocated to it (Fischer et al., 2006). Due to rapid industrialization, urbanization and high population rate increase (up to 3.6%), economic realities seem certain to reallocate water increasingly away from agriculture to other sectors though the demand for more food and fibre is increasing steadily. Therefore, it is necessary to either develop new techniques for conserving water under rainfed agriculture or modernise the existing practices. Similarly, innovative irrigation techniques need to be improved upon in order to ensure optimal use of allocated water and at the same time justify the investment in the sector.

Irrigated agriculture is facing new challenges in the present time. Formerly irrigators focused on design. However, the current challenges include water scarcity, competing water users, cost of implementing irrigation projects, water quality and efficient water usage. Following the recent downward trend in freshwater allocation to agriculture, the sector is under heavy pressure to produce more food in order to meet the demands of the increasing population. This will amount to increasing Crop Water Productivity (CWP) that is, increasing the benefits that are derived from the use of water in crop cultivation under both rainfed and irrigated agriculture. In technical terms, CWP is the ratio of economic crop yield to the amount or depth of water used in producing it (Kirda et al., 1999; Molden et al., 2003). In order to meet these new challenges, a more precise technique needs to be incorporated into the existing methods of irrigation scheduling in order to effectively manage water resources. These include new and efficient designs of irrigation systems, innovations and management of existing facilities in order to ensure sustainability and adaptability.

2.2   Food security and irrigated agriculture

Irrigated agriculture is very vital in meeting the food and fibre needs of the rapidly increasing human population. Agriculture is the largest consumer of freshwater with an estimated 1,300 m^3 year^{-1} required to produce an adequate diet (Falkenmark and Rockstrom, 2004). Scenario analysis revealed that about 7,100 km^3 year^{-1} are consumed globally to produce crop of which 5,500 km^3 year^{-1} are used in rainfed agriculture and 1,600 km^3 year^{-1} in irrigated agriculture (de Fraiture et al., 2007). Analysis also describes large increases in the amount of water needed to produce food by 2050, ranging from 8,500 to 11,000 km^3 year^{-1}, depending on assumptions regarding improvements in rainfed and irrigated agricultural systems. It has been stated in different fora that there is an urgent need for agriculture to scale up its food production with less water for the world population, which is on the increase on daily basis in order to reduce poverty and hunger (Howell, 2001; Food and Agriculture Organization of the United Nations (FAO), 2009). Over the recent decades there has been a steady increase in irrigated lands of about 40,000 km^2 year^{-1} (Ararso, 2005). The irrigated land area in 1970 was 1.69 million km^2 when the world population was 4.1 billion but increased to 2.11 million km^2 in 1980 when the population was 4.9 billion. From 1990 to 2000, there was progressive increase in irrigated land area from 2.39 million km^2 to 2.78 km^2 when the population increased from 5.6 billion to 6.1 billion respectively with some 1.80 million km^2 provided with drainage (International Commission on Irrigation and Drainage (ICID), 2008).

At the end of 2005, there were about 6.5 billion people in the world. About 85% of them lived in the emerging and the least developed countries with an average growth rate of 1.2% per year while the others lived in the developed countries with a growth rate of 0.6% per year. Similarly, the world population density of arable land is expected to increase from the current 430 to 525 persons km^{-2} in 2025 and 600 persons km^{-2} in 2050 (Schultz et al., 2005). In 2005, Asia had the highest population density of 701 persons km^{-2} followed by the least developed countries (520 persons km^{-2}) and the emerging countries (484 persons km^{-2}). The world population is expected to grow to some 8 billion in 2025 and 9 billion in 2050. The expected increase in population and standard of living will result in decline in arable land per capita, decline in annual renewable water resources per capita, increase in need for food production and competition for fresh water among different sectors. Statistics show that currently only 8% (2% with drainage and 6% with irrigation) of the total arable land in Africa is equipped with water management systems (International Commission on Irrigation and Drainage (ICID), 2006).

2.3   Challenges of irrigated and rainfed agriculture

The current challenges facing agricultural water management are different from what they were few decades ago. In the previous years, consideration was not generally given to limitations of the available water resources and supply during irrigation projects planning. The design of irrigation schemes did not address situations in which moisture availability is the major constraint on crop yields. However, in arid and semi-arid regions, increasing rural, municipal and industrial demands for water are necessitating major changes in irrigation management and scheduling in order to increase the CWP in agriculture. Many agronomic and water conservation measures can reduce considerably the amount of water required to produce a crop. These include varying tillage practices, mulching and the use of anti-transpirants. More than one third of the global food supply is produced through irrigated agriculture (UN-WWAP, 2006). Globally, there are sufficient land and water resources to produce food for the next 50 years if the water resources are well managed. The present water scarcity at local and regional scales will hinder effort to increase food production in major agrarian regions and communities. It has been estimated that about 900 million people live in water scarce river basins (closed basin) while another 700 million people live in areas where access to water resources is fast approaching. Worse, still another 1 billion people live in basins where economic constraints limit the pace of needed investment in water management (Molden et al., 2007). The production of sufficient food in order to meet the future food needs requires water development and management strategies that promote improvement in food security and at the same time maintain productivity of our land and water resources and enhances social and environmental amenities. Without increment in productivity, an additional 5,000 km^3 will be required to meet future food demands (de Fraiture et al., 2007). There are various strategies with varying success in managing water efficiently and achieve food and livelihood security. These include increasing the existing agricultural land, avoiding expansion of low-productivity agriculture and improvement in CWP (Yang and Zehnder, 2007). Expanding land may be a good option where the resources, such as finance and land are available. However, in Nigeria where population is expanding at a fast rate and competition for land has become a critical issue, more result oriented and sustainable measures need to be taken in order to manage land and water resources without further stressing the water-limited system.

Ogun-Osun River Basin is one of the river basins in Nigeria. It covers the sub-humid ecological zone of the south-western states of Nigeria. The challenges in the basin are similar to those outlined before. In the basin, there is the need to use the land and water resources for production of food and cash crops on a sustainable and productive basis. Since more than 90% of the crop production in the basin is rainfed, major efforts need to be made to increase productivity by introducing effective soil and water conservation measures during the rainy seasons. Similarly, productivity during dry season farming can also be improved by using low gravity drip irrigation systems.

2.4   General profile of Nigeria and geographic location

Nigeria, the tenth largest country in Africa lies on the west coast of Africa and occupies a land area of 923,768 km^2 including about 13,000 km^2 of inland water (FAO, 2010c). Nigeria is located on the latitude 4° N and 14° N and longitude 2° 2' E and 14° 30' E. The country's North-south extent is about 1,050 km and its maximum East-West extent is about 1,150 km. Nigeria is bordered to the West by Benin, to the Northwest and North by Niger, to the Northeast by Chad and to the East by Cameroon, while the Atlantic Ocean forms the southern limits of the Nigerian territory (Figure 2.1). Land cover ranges from thick mangrove forests and dense rain forests in the South to a near-desert condition in the north-eastern corner of the country. In Nigeria, three broad ecological zones are commonly distinguished and these are: (i) The Northern Sudan Savannah; (ii) The Guinea savannah zone or Middle Belt; (iii) The Southern rainforest zone. Based on meteorological data such as rainfall and temperature, Nigeria is divided into eight agro-ecological zones (Table 2.1) (FAO, 2010c).

Table 2.1. Agro-ecological zones in Nigeria (Food and Agriculture Organization of the United Nations (FAO), 2010c)

Zone	Percentage of country area (%)	Annual rainfall (mm)	Monthly temperature (° C)		
			Minimum	Normal	Maximum
Semi-arid	4	400 - 600	13	32 - 33	40
Dry sub-humid	27	600 - 1 000	12	21 - 31	49
Sub-humid	26	1 000 - 1 300	14	23 - 30	37
Humid	21	1 100 - 1 400	18	26 - 30	37
Very humid	14	1 120 - 2 000	21	24 - 28	37
Ultra humid (flood)	2	> 2 000	23	25 - 28	33
Mountainous	4	1 400 - 2 000	5	14 - 29	32
Plateau	2	1 400 - 1 500	4	20 - 24	36

Nigeria's coastline along the Gulf of Guinea totals 853 km. Nigeria has a territorial sea of 22 km, an exclusive economic zone of 370 km, and a continental shelf to a depth of 200 m. The country has five major geographic regions. These are a low coastal zone along the Gulf of Guinea; hills and low plateaus north of the coastal zone; the Niger-Benue River Valley; a broad stepped plateau stretching to the northern border with elevations exceeding 1,200 m+MSL (mean sea level); and a mountainous zone along the eastern border, which includes the country's highest point, Chappal Waddi (2,419 m+MSL). Nigeria has two principal river systems: the Niger River and Benue River. Niger River is the largest in West Africa, flows 4,000 km from Guinea through Mali, Niger, Benin and Nigeria before it discharges into the Gulf of Guinea (Figure 2.1). Benue River, the largest tributary, flows 1,400 km from Cameroon into Nigeria, where it discharges into Niger River. The country's other river systems involve various rivers that merge into the Yobe River, which then flows along the border with Niger and discharges into Lake Chad according to Library of Congress (FAO, 2010c).

2.4.1   Climate

Nigeria climate is semi-arid in the North and humid in the South. Except for an ultra-humid strip along the coast with rainfall averages of over 2,000 mm year^{-1}, where it rains almost all year round. Rainfall patterns are marked by distinct wet and dry seasons. Rainfall is concentrated in the period June to September. Deficiency in total annual precipitation occurs in the Northern parts of the country. In most other areas, however, the rainfall distribution varies in time and space even with low dependability. Mean annual rainfall over the whole country is estimated at 1,150 mm. It is about 1,000 mm in the inter land (centre of the country) and 500 mm in the Northeast. Mean annual pan evaporation is 2,450 mm in the Southeast, 2,620 mm in the Inter-land and 5,220 mm in the northern part of the country. Total cultivable area is estimated at 610,000 km^2, constituting about 66% of the total area of the country. In 2002, the cultivated area was 330,000 km^2, of which arable land covered 302,000 km^2 and permanent crops 28,000 km^2. About two-thirds of the cropped area is in the North, while the rest are about equally distributed between the Middle Belt and the South. Daily air temperatures range between 13 - 40 °C in the semi-arid North and 18 - 37 °C in the humid South (Table 2.2). High average humidity of 81% is normally observed from February to November in the South and about 40% from June to September in the North. This low humidity coincides with the dry season. The average land areas allocated to agricultural in Nigeria from 1961 to 2000 are shown in Table 2.2.

Table 2.2. Average agricultural land area in Nigeria (1961-2011)
(Oyekale, 2007; FAO, 2013)

Period	Average agricultural land area (10^6 ha)	Growth rate (%)
1961-1965	69.0	0.15
1966-1970	69.5	0.23
1971-1975	69.9	0.04
1976-1980	70.3	0.10
1981-1985	70.7	0.23
1986-1990	71.7	0.25
1991-1995	72.6	0.17
1996-2000	71.3	0.23
2001-2005	74.1	1.53
2006-2011	76.6	-0.49

There was progressive increase in agricultural land use from 1961-1970. From 1970-1990, the land area subjected to crop production reduced drastically and this can be attributed to oil boom and low level of investment in the development of agricultural land. The growth rate in agricultural land area reduced drastically from 1991-2000. Reduction in agricultural land area in the recent times could be attributed to rural urban migration, low level of investment in crop production by government and over dependence on oil revenue. Despite the sharp reduction in the growth rate from 2006 to 2011, the population keeps increasing and there are more mouths to feed than before. In the light of this, increasing the CRP that is, more crops per drop at basin scale will go a long way in ensuring food sustainability in Nigeria in order to feed her teeming population. This is the focus of this study.

2.4.2 Administrative divisions

Nigeria is made up of 36 states and Abuja is the Federal Capital Territory (FCT). These are further divided into 774 local government areas. It is divided into six geo-political zones for effective administration and equitable sharing of the Nations' resources for social and economic development (Figure 2.1). There are more than 350 ethnic/linguistic groups in Nigeria and a variety of social groups. In 2002, 60% of the total population was using improved drinking water sources, with 72% in urban areas and 49% in rural areas (FAO, 2010c). There were high levels of poverty during the 1980s and 1990s, with more than 35% of the population living below the US$ 1 per day poverty level in 2001. Poverty is very rampant in rural areas, where 40% of the population lives below the poverty line. More than 5% of the rural dwellers are suffering from HIV/AIDS and more than 50 million Nigerians suffer from a combination of diseases of protein-energy malnutrition.

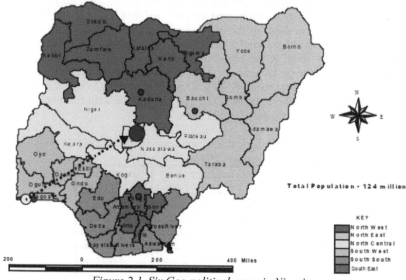

Figure 2.1. Six Geo-political zones in Nigeria

Nigeria is by far the most populous country in Africa, with its 127 million people accounting for about one-seventh of the total population of Africa's 53 countries. Population density is 138 inhabitants km^{-2}, annual growth rate is 2.2% and 52% of the population is rural. In 2006, the population was 148 million while the projected population for 2025, 2050 and 2100 are 240, 440 and 914 million respectively (Figure 2.2). Based on this geometric increase in human population, there is a need to scale up food production with minimum pressure on available freshwater resources. This can be achieved by intensifying efforts on sustainable use of land and water resources under rainfed conditions and practices of deficit or regulated irrigation in dry season farming.

2.4.3 Economy

Nigeria is an emerging economy whose revenue largely depends on crude oil, which accounts for about 90% of total exports and for about 70% of government revenues. The country's Gross Domestic Product (GDP) in 2003 was estimated at US$ 50.2 billion, and in 2002 the contribution from agriculture was 37.4%, with about 90% of the agricultural output coming from subsistence farming. Agriculture provides occupation for 30% of the economically active population. 38% of agricultural workers are female while 62% are male. The FAO has listed Nigeria among those nations that are nowadays technically unable to meet their food needs from rainfed production at a low level of inputs and appear likely to remain so even at intermediate levels of inputs at some points in time between 2000 and 2025. Farming is still being practiced at subsistence level byusing crude tools, and agricultural landholdings are scattered. Simple, low-input technology is employed, resulting in low-output labour productivity. Typical farm sizes range from 0.5 ha in the densely populated high-rainfall South to 4 ha in the dry North. Nigeria faces immense challenges in accelerating growth, reducing poverty and meeting the Millennium Development Goals (MDG). In May 2004, Nigeria launched its National and State Economic Empowerment and Development Strategies (NEEDS and SEEDS) for growth and poverty reduction. NEEDS is based on three pillars: (i) empowering people and improving social service delivery; (ii) improving the private sector and focusing on non-oil growth; (iii) changing the way government works and improving governance. Some good progress has been made, particularly at Federal level on macro-economic stabilization, fuel subsidies and procurement. However, much remains to be done, especially at the local level where the implementation of SEEDS is proving more difficult than its Federal counterpart is as expressed in National Millennium Development Goals (NMDG) (2004).

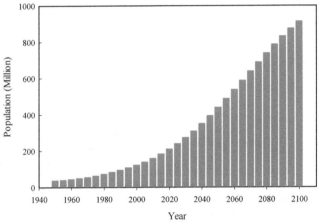

Figure 2.2. Population trend in Nigeria since 1950 and the projections (medium variant) for the future (World Population Prospects (WPP), 2012)

2.5   Water resources management and state of use in Nigeria

Nigeria is well drained with a close network of rivers and streams. Some of these, particularly the smaller ones in the North, are seasonal. There are four principal surface water basins in Nigeria namely:

- the Niger Basin, which has an area of 584,193 km^2 within the country. It constitutes 63% of the total area of the country and covers a large area in Central and North-west Nigeria (Figure 2.3). The key rivers in the basin are the Niger and its tributaries Benue, Sokoto and Kaduna;
- the Lake Chad Basin in the Northeast with an area of 179,282 km^2, or 20% of the total area of the country. It is the only internal drainage basin in Nigeria. Important rivers are the Komadougou Yobe and its tributaries Hadejia, Jama'are and Komadougou Gena;
- the south-western littoral basins have an area of 101,802 km^2, which is 11% of the total area of the country. The rivers originate in the hilly areas to the south and west of the Niger River;

- the south-eastern littoral basins, with the major watercourses being the Cross and Imo rivers, have an area of 58,493 km², which is 6% of the total area of the country. They receive much of their runoff from the plateau and mountain areas along the Cameroon border (FAO, 2010c).

Figure 2.3. River basins in Nigeria (FAO, 2005)

In Nigeria, water resources management at basin scale is vested into the River Basin Authority. Currently, there are twelve river basin authorities established for the purpose of managing water use and allocation (FAO, 2005). These are (RBDAN, 1979):
- Sokoto-Rima River Basin Development Authority;
- Hadejia Jama're River Basin Development Authority;
- Chad Basin Development Authority;
- Upper Benue River Basin Development Authority;
- Lower Benue River Basin Development Authority;
- Cross River Basin Development Authority;
- Anambra River Basin Development Authority;
- River Niger Basin Development Authority;
- Ogun-Osun River Basin Development Authority;
- Benin-Owena River Basin Development Authority;
- Niger Delta Basin Development Authority.

The total cultivable land area in Nigeria was estimated at 61,000 km², which constitutes about 66% of the total area of the country. The cultivated area was 3.3 million km² in 2002, of which arable land covered 3.02 million km² and permanent crops 28,000 km². About two-third of the cropped area is in the far North with the rest about equally shared between the Middle belt and the South. The total water withdrawal in Nigeria in 2000 was 8.004 km³ and agriculture accounted for about 68.8% of the withdrawal (Figure 2.4).

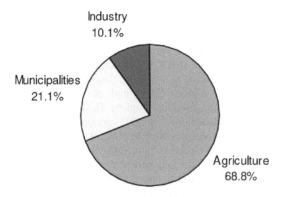

Figure 2.4. Water withdrawal in Nigeria (FAO, 2005)

In 2004, the total land area equipped for irrigation was 2,931 km² but the total area actually irrigated was 2,188 km² with about 30% of the population economically active in agricultural activities. Different arms of government are involved in irrigation. The structure of involvement is shown in Figure 2.5.

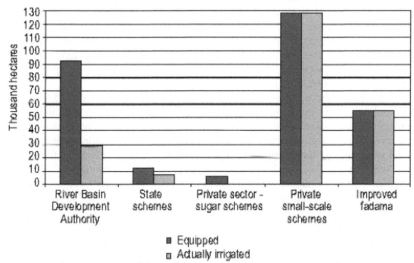

Figure 2.5. Structure of the irrigation sector in Nigeria in 2004 (FAO, 2005)

In 2008, the population in Nigeria was 152 million with a rural population of 78 million people, but the population that is economically active in agricultural had reduced to about 8% without remarkable improvement in the use of modern farming techniques (FAO, 2010a). The output in terms of annual agricultural productivity, such as crop yield from the various agrarian communities cannot justify the land and water resources for which the Ogun-Osun River Basin is endowed with. This can be directly attributed to inadequate planning and management of land and water resources, especially for food production (Adekalu et al., 2002; Adeboye and Alatise, 2008).

Based on available statistics agriculture plays prominent roles in the economic development and poverty reduction of Nigeria. Agriculture accounts for about 35% of the Gross Domestic Product (GDP) in Sub-Saharan Africa and engages about 70% of the human population either directly or indirectly (Irz and Roe, 2000; World Bank (WB), 2000 and 2005).

In order to achieve total eradication of poverty and hunger, a major part of the Millennium Development Goals (MDG), substantial growth in agricultural output is essential. Precisely, food production has to be doubled over the coming 25-30 years, especially in Sub-Saharan Africa (SSA) and parts of South East Asia where malnutrition and demand for food are highest (UN Millennium Project (UNMP), 2005a,b). Irrigation is an adaptation to rainfall variability either on seasonal or annual basis. The practice of irrigation dates back to about 4,000 years (Framji et al., 1982) in both the humid tropics where rainfall is high and in the arid and semi arid mid latitudes where rainfall is too little, erratic and poorly distributed to sustain food production (Postel, 1999). The contributions of irrigated agriculture to social economic well being of any place where it is being practiced cannot be overemphasized. Despite the fact that the management of investment in irrigated agriculture has been questioned at many occasions (Repetto, 1986) irrigated agriculture has been thought to have made a massive contribution to the global food security (Molden et al., 2007) and also to improve livelihood (Lipton et al., 2003). On a global scale, irrigated agriculture provides approximately 40% of the world's food from less than 20% of its cultivable area (FAO, 2005). Worldwide, some 2.70 million km^2 land area are irrigated, accounting for two-third of the total water consumption, which is currently estimated at 4 billion m^3 $year^{-1}$. In the emerging and least developed countries, the proportion of water used for irrigation is often higher, at 75-80% of the total use (Perry, 2007). The challenges facing agricultural water management today are very different from what they used to be in the time past. Currently the global population is on the increase and the standard of living of the people has changed drastically especially in the developed and emerging countries (Schultz et al., 2005).

In Nigeria, the population increased from 89 million in 1986 to 150 million in 2009 and by 2025, the projected population of Nigeria is 217 million (National Population Commission of Nigeria (NPCN), 2010). The implication is that there are more mouths to be fed, and more land areas need to be developed in order to meet the current and future food challenges. In Ogun-Osun River Basin in Nigeria, the irrigation and drainage projects established in the 1960s by the then government of the Western Region have been abandoned. The current energy crisis in Nigeria has made it difficult to use highly sophisticated and capital intensive irrigation systems during the dry season in Ogun-Osun River Basin and other river basins in Nigeria. Irrigation equipment procured, such as sprinkler irrigation systems, are in a state of disrepair for a long period of time. Soil assessments carried out revealed that 176,544 km^2 of the mapped 255,167 km^2 in Nigeria are severely degraded (FAO, 2010c) and Ogun-Osun River Basin constitutes about 65% of the land area that has been described as severely degraded (section 3.2). Currently, about 24 million people are living within Ogun-Osun River Basin and a substantial proportion is experiencing inadequate water supply in terms of quantity and quality similar to other locations in Nigeria. Consequently, the land and water resources in this basin have remained overexploited for social and economic benefits of the immediate communities and the entire Nigerians for a long period of time (Alatise and Adeboye, 2005). In addition, climate change has resulted in fluctuation in the rainfall pattern in the basin over the years. For instance, in 2012 there was no rainfall until May, while in 2012 no significant rainfall was recorded until June. This is contrary to the previous commencement of rainfall as early as the month of March. Therefore, there is an urgent need to manage the existing land and water resources on a productive and sustainable basis in the basin.

Soybean has been found as one of the promising crops, which can serve this purpose. If water and land resources in the Ogun-Osun River Basin are utilized for the cultivation of Soybeans, the basin is expected to produce a substantial quantity of the crop, which will meet both the local and national needs. Similarly, jobs will also be created for the teeming population of able and agile retirees and unemployed graduates in Nigeria. Land resources and nutrients, which may be depleted after intensive crop cultivation, will be naturally replenished if cultivated in the rainy season under water conservation practices and deficit irrigation in the dry season. Soybeans have a high capacity to naturally fix nitrogen in the soil. Therefore the focus of this study is to determine the water use pattern of Soybeans under both rainfed and deficit irrigated agriculture in Ile-Ife and to provide information on how to ensure productive and sustainable use of land and water in Ogun-Osun River basin under rainfed and deficit irrigated agriculture.

2.6   Problem description

Human population is increasing faster than the available food and freshwater resources globally. The current fresh water allocation of about 72% to agriculture is decreasing because of competing users from other sectors, such as industries and municipalities (Cai and Rosegrant, 2003). There is an urgent need to scale up agricultural production in order to meet the needs of both industries and domestic users, especially in less developed countries and poverty-stricken regions, such as Sub-Saharan Africa and Latin America (Howell, 2001; Molden et al., 2007).

By 2025, the projected human population in Nigeria is 217 million. Out of the current 150 million people about 24 million (17%) reside within Ogun-Osun River Basin (Figure 2.6). The basin is densely populated. This indicates that there will be more mouths to be fed in the nearest and distant futures. Therefore there is a need to scale up food production on a sustainable basis in the basin. Scaling up of production can be achieved by increasing CWP of food crops (Kijne et al., 2003; Zwart and Bastiaanseen, 2004) and through careful management of water resources under both rainfed and irrigated agriculture (deficit irrigation). Increasing crop water productivity means more crops should be produced per every drop of water used in the basin (Nigerian National Committee on Irrigation and Drainage (NINCID), 1999).

The government of the Western Region invested many resources in the cultivation of crops such as Cocoa, Maize, Palm fruit in Ogun-Osun River Basin in the 1970s and 1980s. Today the productivity of these cash crops has reduced substantially because of intensive cultivation of these crops and have led to reduction in soil fertility. A large proportion of the basin including some other areas in Nigeria has been described as severely degraded. Rainfall fluctuations in space and time in recent times in the basin indicate that pragmatic measures of adaptation need to be taken. In order to ensure productive and sustainable use of land and water in the basin. It is thereby imperative to cultivate alternative crops that will replenish soil nutrients and to study management practices that will ensure optimum use of water in the soil. If these are done, land resources will be conserved and a regular income for the farmers will be maintained (sustainability). One of such crops is Soybean. Soybean is cultivated in Nigeria, especially during the rainy season. In the dry season, however, cultivation of Soybean is reduced due to shortage in water supply. Aside from transferring a fixed amount of Nitrogen to the inter-planted crops, Soybean has the ability to bring minerals from deeper soil horizons to the surface thereby improving soil air circulation to the benefit of the accompanying crop. Soybean is relatively high nutritious and hence has a wide acceptance in terms of cultivation among the small-scale peasant farmers in Nigeria. It is also used as supplement to the traditional cereal, or tuber based diets of most Africans who generally are known to be suffering from protein deficiency. Hence, consumption of Soybean is generally recommended in various school feeding programs and to the mal-nourished populations in developing countries such as Nigeria.

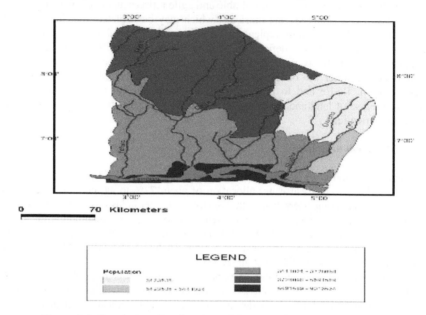

Figure 2.6. Population distribution in Ogun-Osun River Basin, Nigeria
(Federal Ministry of Environment Nigeria (FMEN), 2010)

More than three quarter of the national agricultural land area is rainfed. This is attributed to many factors such as adequate rainfall in many parts of the country during the rainy season. Lack of affordable irrigation facilities that can be used for dry season farming by peasant farmers is another factor (NINCID, 1999). Although rainfall patterns vary across agro-ecological areas and the yields are directly connected to this variability. Increasing, the potential of rainfed agriculture would make a significant impact on the food production in Nigeria (NINCID, 1999). In order to achieve this, there is a need to develop and implement intermediate low cost water conservation practices, which local farmers - who are the major food producers - can adopt in cultivating their crops in the rainy season. The technologies include water harvesting, soil tillage, mulching, bunding and terracing. If well managed and sustained the use of these technologies could significantly reduce the water risk under climate change and lead to substantial increase in the yield of crops in Ogun-Osun River Basin.

Several cultivars of Soybeans that are well adaptable to the local conditions in different parts of Nigeria have been produced at the International Institute for Tropical Agriculture (IITA) in collaboration with many research institutes in West Africa (Tefera, 2011). Effort should not be limited to the production of new cultivars only. There is a need for research on how to ensure sustainability in their production under the present climate change. Currently, there is no scientific information on the response of those cultivars of Soybean to deficit irrigation in the sub-humid and humid regions of Nigeria. Similarly, under the current fluctuation in rainfall and reduction of soil moisture, there is a need to introduce innovative cultivation practices that will ensure the production of the crop on a continuous basis. Through field experiments, the proposed research will furnish on-farm information on the response of Soybeans to different water application and deficit irrigation using the in-line drip irrigation system. Similarly, the possibility of improving the yield and WP of the crop by using soil and water conservation techniques will be examined. Subsequently, enhanced cultivation of Soybeans in the basin will foster productive and sustainable use of land and water resources for food production in Ogun-Osun River Basin and Nigeria at large.

2.7   Research questions

It is evident from the available statistics that the fresh water allocation to agriculture is reducing due to population pressure and urbanization, which are more rampant in the least developed and emerging countries of the world. Sub-Saharan Africa is a region with more than 150 million undernourished people and is facing environmental challenges such as land degradation and acute physical and economic shortages of fresh water. In the same context, the main research question to be addressed in this study has been formulated as follows.

'To what extent can the land and WP of Soybeans be sustainably increased under water conservation practices, full and varied irrigation conditions in Ogun-Osun River Basin, Nigeria?'

This raises the following specific questions:
- how does seasonal and spatial variation of rainfall affect yield of Soybeans in Ogun-Osun River Basin?
- how does water conservation affects soil water storage and temperature under rainfed conditions?
- how does water conservation affects components of the soil water balance such as transpiration, evaporation and seasonal crop water use of Soybeans?
- does seasonal rainfall variability affects WP and harvest index (HI) of Soybeans?
- what is the effect of deficit irrigation on WP for Soybeans?
- to what extent will skipping of irrigation every other week during reproductive stages reduce the seed yields of Soybeans?
- to what extent will deficit irrigation of Soybeans help increase or maintain production of the crop while conserving water and land in the basin?
- how will the cultivation of Soybeans under varied irrigation conditions contribute to sustainable use of land and water in Ogun-Osun River Basin?

2.8   Hypothesis

Poor management of water and land resources for crop cultivation and land degradation are major challenges in Ogun-Osun River Basin and in other agro-ecological zones in Nigeria. Urbanization, changes in life style and increasing human population are mounting more pressure on the water and land resources in the Ogun-Osun River Basin on daily basis. Similarly, climate change characterized by fluctuation in rainfall in time and space requires pragmatic strategies in order to adapt to these challenges. It is hypothesized here that given the available water resources, human population and land use practices in Ogun-Osun River Basin, the land and water resources can be optimally and sustainably utilized by adopting innovative and adaptive technologies of water management and agricultural practices under rainfed conditions and deficit irrigation. Among the innovative methods are water conservation at field level and cultivation of crops that replenish soil nutrients with little or no reliance on artificial fertilizer.

2.9 Expected outcomes and contributions to knowledge

The study is expected to provide insights and increased understanding:
- of the effects of water conservation practices on water storage in the soil and yield components of Soybeans;
- of seasonal variability of canopy structures and utilization of solar radiation (SR) for biomass accumulation;
- of stakeholders in the agriculture sector such as farmer unions, government both at federal and state levels on the prospects of cultivating Soybeans under deficit irrigation by using drip irrigation;
- on the contribution to the global effort in conserving fresh water;
- for developing a crop yield model for Soybeans and similar crops under water deficit conditions;
- that will serve in the preparation of guidelines for effective management of land and water for cultivation of Soybeans in Ogun-Osun River Basin.

2.10 Research objectives

The research objectives of the study are to:
- determine the extinction coefficient, fractions of Intercepted Photosynthetically Active Radiation (fIPAR), dry matter and Radiation Use Efficiency (RUE) of Soybeans and their seasonal variability under rainfed conditions;
- determine the seasonal water use, WP of Soybeans under rainfed agriculture;
- determine the effects of water conservation practices on water storages in the soil and their impacts on yields, seasonal transpiration and evaporation, HI and transpiration efficiency of Soybeans;
- generate regression models and relationships between yield, WP, IPAR and RUE;
- determine the effects of water stress on biometric data such as plant heights, number of leaves, leaf area index (LAI), yield, water use, WP and IWP of Soybeans and the stage of growth in which yield and yield components of the crop can be greatly reduced due to deficit irrigation;
- use of the AquaCrop model to simulate canopy cover (CC), water storage in the soil, dry above ground biomass (DAB), yield and WP of the crop by using field data under different irrigation conditions;
- validate the use of the AquaCrop model in predicting response of Soybeans to full and deficit water application at different phenologic stages;
- recommend agronomic and water saving measures that can ensure productive and sustainable use of land and water resources for Soybean cultivation in Ogun-Osun River Basin.

3 Ogun-Osun River Basin

Ogun-Osun River Basin is located in the South Western part of Nigeria, latitudes $8° 20'$ N and $6° 30'$ N and longitudes $5° 10'$ W and $3° 25'$ W (Figure 3.1) with a land area of 101,802 km^2, which is 11% of the total area of the country. Land cultivation and farming activities are carried out in almost all parts of the basin and this led to the establishment of farm settlement schemes in strategic locations within the basin. Food and cash crops such as Cocoa, Kola, Palm trees, Plantain, Maize, Yam, Cocoyam, etc. are planted in the basin. The river basin covers the present Lagos, Ogun, Osun, Oyo and parts of Kwara states of Nigeria. The creation of the River Basin and Rural Development Authorities (RBRDA) in Nigeria was motivated by the desire of the Federal Government to facilitate and accelerate food production to cater for the demands of the teeming population and to open up rural settlements throughout the country for increased food production (River Basin Development Authority of Nigeria (RBDAN), 1979).

The inability of RBRDAs to perform their functions effectively can be attributed to the factors of political environment and actions of political actors, which have been incompatible with managerial and organizational goals of the authorities since their establishment (Akindele and Adebo, 2004). The land use pattern in Ogun-Osun River Basin is shown in Figure 3.2. A larger proportion of land in the basin is used under intensive (row crop, minor grazing) smallholder rainfed agriculture. In the sections that follow, the climate, soil and land resources of the basin will be explained.

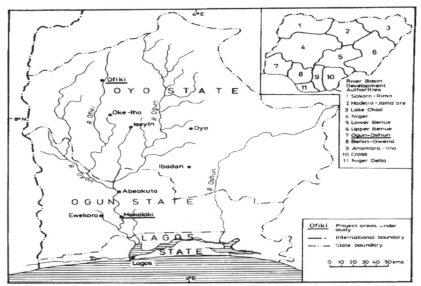

Figure 3.1. Ogun-Osun River Basin and the locations of other River Basins in Nigeria (adapted from Areola et al., 1985)

3.1 Climate

Generally, the Ogun-Osun River Basin's climate is influenced by the movement of the inter-tropical convergence zone (ITCZ), a quasi-stationary boundary zone, which separates the sub-tropical continental air mass over the Sahara and the equatorial maritime air mass over the Atlantic Ocean. The former air mass is characterised by the dry North-Easterly winds called the Barmattan found in the rain-bearing South-Westerly winds from the Gulf of Guinea as reported by Adeboye (2005).

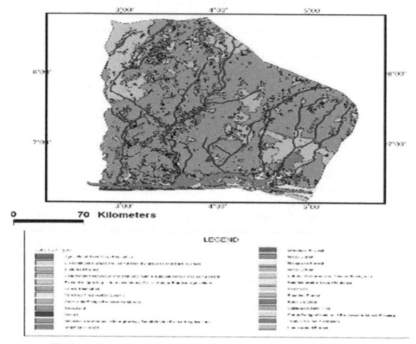

Figure 3.2. Land use pattern in Ogun-Osun River Basin, Nigeria

3.2   Rainfall and air temperature

Seasonal distribution is the main feature of the rainfall pattern in Ogun-Osun River Basin. The mean annual rainfall in Nigeria is shown in Figure 3.3. The rainy season emerges earlier in the South in March and continues until end of October or early November, with at least seven months of rainfall. In the north of Ogbomoso the rain starts in early May or late April and ends in the middle of October. Dry days are regular and sufficiently regular in late July and early August to constitute a 'little dry season' whose monthly precipitation depth is below 120 mm. In the wet season, the mean rainfall ranges between 1,020 and 1,520 mm in the south of the basin, but in the north, it is less than 1,020 mm. In the North and South, the mean dry season rainfall varies from 127 to 178 mm and 178 to 254 mm respectively.

The record of temperature in the basin shows that the hottest months are February and March during which temperatures are high over the entire area. For the month of February, the mean daily maximum temperature is 31.4 ^{0}C in the North. The minimum recorded temperature during Hammatan in the North is 47 ^{0}C. During the rainy season in July, a lowest mean minimum temperature of about 22.8 ^{0}C was recorded (Ogun-Osun River Basin and Rural Development Authourity (OORBA), 1982).

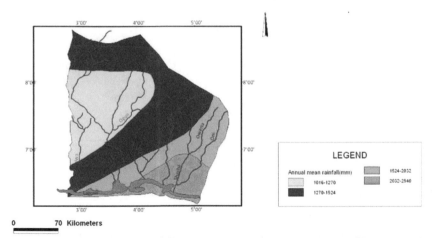

0   70   Kilometers

Figure 3.3. Mean annual rainfall in Ogun-Osun River Basin, Nigeria (FMEN, 2010)

3.3   Soil and land resources

Due to the erosion and sediment transport, the soils in the basin have developed into alluvial parent materials. The basement complex in the upper part of the basin gives rise to a wide variety of soils, coarse in texture and of low fertility. In Ogun-Osun River Basin intensive (row crops, minor grazing) smallholder rainfed agriculture dominates (Figure 3.4). Undisturbed forest is scanty because of the frequent lumbering activities in the basin over the years. For productive intensive crop production in the basin, irrigation and heavy fertilizer application is highly required. The soils in the basin are classified into two groups based on location and elevation (OORBA, 1982). These are:

- the upland soils, which are more developed and range from heavy and hydromorphic to coarse and well-drained (Figure 3.4);
- the lowland soils, which are hydromorphic and affected by a high groundwater table and seasonal flooding.

Water resources in Ogun-Osun River Basin include surface water and groundwater. Surface water plays a prominent role in the basin. Sometimes limited streamflow records create problems in water resources assessment. Generation of long record is carried out by streamflow synthesis of rainfall records. In Ogun-Osun River Basin, the two major potential sources of undergroundwater are the coastal plains sand, incorporating the upper part of Ilaro and Abeokuta formations. These formations have the following common features:

- the origin of the deposition;
- the mode of deposition.

The movement of the groundwater is from North to South towards the sea. The aquifer units to East and West are assumed to coincide with the boundaries of the surface river basins. The formation in Abeokuta comprises of the phreatic zone, which is replenished by percolation of infiltrated rain water through the unsaturated formation and the confined zone. It dips to the South of the formation outcrops with area considerably in excess of the unconfined phreatic zone. Due to the relative abundance of surface water in the basin, groundwater seems not to be of significance for crop cultivation and consumption (OORBA, 1982).

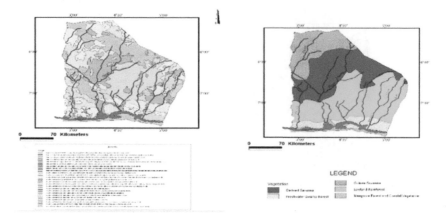

Figure 3.4. Soil and vegetation patterns in Ogun-Osun River Basin (Ogun-Osun River Basin and Rural Development Authority (OORBA), 1982)

3.4   Current challenges in Ogun-Osun River Basin

Irrigation and drainage projects established in the 1960s by the then government of the Western Region have suffered neglect by the previous political administrations. Equipment procured for land and water management such as sprinkler irrigation facilities are in a state of disrepair for long period. In few places where old and poorly maintained sprinkler irrigation facilities are found, the current energy crisis in Nigeria has made it difficult to use them on sustainable basis for crop cultivation during the dry season. Consequently, economic returns on that equipment have not justified the huge investment made on them years ago. The use of sprinkler irrigation in wetting crops does not ensure conservative use of water at this time when there is increasing need to optimise the use of water at farm scale. Therefore there is a need to use drip irrigation which concentrates water at the root zone of the crops and optimises the use of water under both supplemental and deficit irrigation conditions.

Various schemes established to maximise the use of water for crop production in the basin have been either abandoned or not given appropriate attention by the agencies whose responsibility is to manage them effectively to attain regional and national sufficiency in crop production. Soil assessments carried out revealed that 177,000 km^2 of the mapped 255,000 km^2 in Nigeria are severely degraded (FAO, 2010c) and Ogun-Osun River Basin constitutes about 45% of the land area that has been described as severely degraded (Figure 3.5). This could be one of the reasons for substantial reduction in agricultural produce in the basin. The increasing human population and daily migration of people to urban centres and cities in the basin call for the need to maximise the production of food crops.

Land and water resources in the basin have remained unexploited for social and economic benefits of the immediate communities and the entire Nigerians for long period of time (Alatise and Adeboye, 2005). Despite the abundant rainfall in certain months of the years in the basin, variability of rainfall in the recent times put crop production at risk. Harvesting of rainfall on the field by using water conservation practices may go a long way in reducing runoff, which eventually leads to loss of soil fertility and land degradation. If water resources are utilized optimally for crop production, Ogun-Osun River Basin will be a major food supplier to markets in West

Africa and Sub-Saharan African countries. There are indications that the attention of the government is shifting from expanding the land for crop production because of the subsidy being paid in accessing other social facilities or services. This has the tendency of reducing production of food crops to feed the teeming population in the basin. Water conservation practices under rainfed conditions have the potential of increasing yields aside of providing enough water for crops and stabilizing the yields under rainfed conditions. The water conservation techniques, which can be practised on the farms in the basin, should be less expensive, environmental friendly and achievable by using materials that are readily available within the environment where farmers live.

Therefore, the focus of this research was to determine the potential of increasing the yield of Soybeans by using water conservation practices under rainfed conditions in Ile-Ife. Similarly, the effects of deficit irrigation on the yield of Soybeans in the dry season has been evaluated. AquaCrop was used in modelling the response of yield to water stress in Soybeans. Results from this study will be communicated to stakeholders in land and water management for implementation in order to ensure sustainable use of land and water resources in the basin.

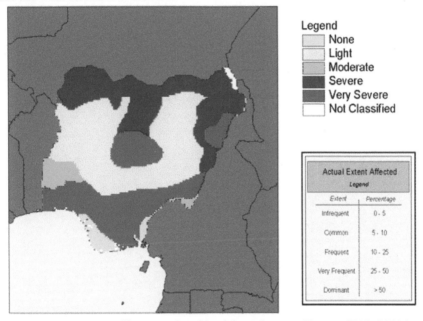

Figure 3.5. Severity of human induced land degradation in Nigeria (FAO, 2010c)

Figure 3.1 ... m

4 Soybean *(Glycine max (L.) Merr.)*

4.1 General botany and descriptions

Soybean is usually erect, bushy annual herb that grows up to 2 m tall, sometimes viny; taproot branched, up to 2 m long, lateral roots spreading horizontally m in the upper 20 cm of the soil; stem brownish or greyish pubescent (Figure 4.1). Soybean seeds have a seed coat and embryo. The seed coat, which is a maternal tissue, has three layers and these are: epidermis, hypodermis, and inner parenchyma layer (Carlson, 1973). On the outside, the seed coat is covered with a cuticle while on the inside there are remnants of the endosperm tissue, which has been compressed by the developing embryo. The seed coat has pores that vary in size, shape and number. These pores form as the seeds desiccate during maturation (Vaughan et al., 1987). There are also surface deposits of a waxy material derived from the endocarp of the pod wall (Yaklich et al., 1986). It was observed that some genotypes have areas of the seed coat without pores, and in others, pores are located over the entire surface. The rate at which Soybean seeds absorb water is dependent on the pore size, distribution and the extent of the waxy deposit on its surface. If pores are smaller or occluded by the waxy material, the seed of Soybeans may not imbibe water at all unless the seed coat is well softened. The seeds of Soybeans, which do not absorb water or take several hours to imbibe, are referred to as hard seeds. Variation in the impermeable response within a hard seed line is associated with seed size, with smaller seeds being more impermeable (Hill et al., 1986). The colours of Soybean seed coats are shades of black, brown, green or yellow as reported by Burton (1997).

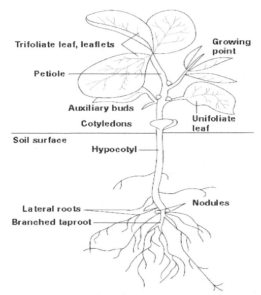

Figure 4.1. A young vegetative Soybean plant (Roth, 2013)

Soybean is a warm season legume and most production occurs in temperate zones of the northern and southern hemispheres. In the northern hemisphere, most planting for a full-season crop is done in May or early June while in the southern hemisphere, full-season crops are planted in November. Soybeans are sometimes planted after the winter crop is harvested, and thus planting may be as late as July or December. In Nigeria, the crop is mostly planted between May and September in the sub-humid agro-ecologic zone, but delay in rainfall may postpone the planting until when the rainfall has stabilized.

4.2 Growth and development

After planting, Soybean seedlings emerge within 5-15 days. A seedbed temperature of 25-33 °C is optimal for the germination of the crop in most climates. The rate of germination depends largely on the soil moisture and temperature. During seed development in the soil, the presence of drought stress reduces seed viability (Dombos et al., 1989; Heatherly, 1993).

4.2.1 Vegetative growth

After germination, the first two leaves to develop above the cotyledonary node are unifoliate. A trifoliate leaf is produced at each node, thereafter it develops on the main stem or branches. Leaflets vary in size and shape but are typically oval or ovate. There is a single recessive gene (ln) which produces a narrow leaflet. Narrow leaflets may allow more light penetration into the crop canopy. Some cultivars have been developed with narrow leaflets. Research conducted to determine whether the narrow leaflet phenotype confers a productive advantage has been inconclusive (Cooper and Waranyuwat, 1985; Wells et al., 1993). Variation occurs among germplasm for leaflet orientation in response to light and temperature (Wofford and Allen, 1982). The colour of the plant varies from dark green to light green depending on the concentration of chlorophyll.

Cultivated Soybeans have an erect growth habit, but procumbency is not uncommon in germplasm sources. The wild Soybeans (Glycine soja) tend to be viny, and many of the Soybean types developed for forage use are viny. Soybeans may have a single main stem with no branching or various degrees of branching and branching response is strongly determined by the plant spacing. More branching occurs at low plant densities. The Soybean root system begins as a taproot developing from the radicle of the germinating seed. Secondary roots in Soybeans develop from the taproot and many orders of branch roots emanate from the secondary roots (Carlson, 1973). Soil and moisture content, cultivation and plant population densities determine rooting depth and prominence of the taproot versus the more fibrous branching roots (Carlson, 1973; Barber, 1978; Robertson et al., 1980).

The vegetative and reproductive periods in the life cycle of Soybean are divided for the purpose of description into sub-stages, which correspond to the appearance in development of main stem nodes and reproductive structures (Fehr and Caviness, 1977). The V1 growth stage is when the unifoliate leaves at node 1 are fully developed and this is followed by V2 when the first trifoliolate leaf has fully developed at node 2. This description continues until the last node (n) develops (stage Vn). The reproductive period is explained from stage R1 (beginning bloom) when at least one flower opens on the main stem, through R4 growth stage (full pod) when a pod of 2 cm long occurs at one of the four uppermost nodes on the main stem, to stage R8 (full maturity) when 95% of the pods have mature pod colour. Both the vegetative and reproductive periods may overlap in certain cultivars, the extent of overlapping is determined by the types of termination of the stems. In Soybean, the types of termination of stem are categorised into three and these are: (i) indeterminate genotypes which tend to have one or two pods near the stem apex; (ii) determinate genotypes which have a well defined terminal raceme with several pods; (iii) semi-determinate types are intermediate between these two in the abruptness with which the main stem terminates (Palmer and Kilen, 1987). In indeterminate stem termination types, the overlap of vegetative and reproductive periods is more pronounced and least in determinate genotypes.

4.2.2 Reproductive growth and maturity

Flowering commences from 25 days to more than 150 days after sowing, depending on day length, temperature, and cultivar (breed). Flowering can take 1-15 days. Soybean is normally self-pollinated and completely self-fertile with less than 1% cross-pollination. Pollen is normally shed in the morning, before the flowers have completely expanded. Flowering and maturity are generally influenced by the photoperiod. This photoperiodic sensitivity limits adaptability of Soybean as a full-season crop to relatively narrow latitudinal belts. The Soybean breeders have developed cultivars with wide adaptation within these areas (Brim, 1973). Soybean is usually weeded 1-3 times during the first 6-8 weeks after planting, after which its canopy should be sufficiently developed to suppress weeds. Irrigation is uncommon in cultivating Soybean except for dry season production. Basal fertilization with 20-25 kg P per ha is often required for adequate symbiotic N_2-fixation and general growth. It is commonly grown in rotation with cereals, such as Maize, Rice, Sorghum, Wheat and Finger millet, whereby all fertilizer may be applied to the cereal (Giller and Dashiel, 2007). After pollination, pods start to develop and are usually visible within five days. Pods reach an almost full width while the developing embryo is still very small. This period between the R1 growth stage (beginning bloom) and R5 growth stage (beginning seed) can vary from 18 to 40 days depending on genotype and environment (Metz et al., 1985). After the R5 growth stage, seeds enter a period of rapid linear dry biomass accumulation. Physiological maturity of the seeds in the pods occurs between the R7 and R8 growth stages when the pods lose their green colour and become yellow (Gbikpi and Crookston, 1981). Pods reach their mature colour mostly (black, brown, or tan) about one week after maturity. Matured seeds are usually harvested after drying in the pod to 15% (or less) moisture. In the humid areas, such as Ile-Ife, the matured seeds are harvested by cutting the stem from the ground and air dried before threshing either manually or mechanical.

4.3   Seed yield

Average world Soybean yields are 2.25 t ha^{-1}; those in the United States are 2.5 t ha^{-1}. Under smallholder farming conditions in tropical Africa yields are often only 0.5 t ha^{-1} due to a combination of poor soil conditions and poor management. However, yields of more than 2 t ha^{-1} have been recorded on smallholder farms in Zimbabwe and Nigeria, particularly when farmers are growing Soybean as cash crop for urban markets or for processing for oil and feed. Under optimal growing conditions, yields of more than 4.5 t ha^{-1} have been recorded in Zimbabwe. In Nigeria and most of West Africa the yield potential of Soybean is about 3 t ha^{-1} (Giller and Dashiel, 2007) (Table 4.1).

4.4   Origin and distribution

Soybean was domesticated in the north-east of China around the 11th century BC. From there, it spread to Manchuria, Korea, Japan and other parts of Asia. The crop was introduced into Korea between 30 BC and 70, and it was mentioned in Japanese literature around 712. It reached Europe before 1737. Soybean was introduced in the United States in 1765 and in Brazil in 1882. It is not clear when Soybean first reached tropical Africa. There are reports of its cultivation in Tanzania in 1907 and in Malawi in 1909, but it is likely that Soybean was introduced during the 19th century by Chinese traders who were active along the east coast of Africa. Currently, Soybean is widely cultivated in tropical, subtropical and temperate regions throughout the world. The slow distribution of the crop outside Asia can be attributed to the absence of Soybean specific rhizobia in the soils of other regions (Boerma and Specht, 2004). Nigeria is one of the countries in tropical Africa where Soybean is cultivated. The available statistics show that the production of Soybeans in Nigeria over the past fifty-one years has been fluctuating especially from 1961 to 1990. In the recent times, however, from 1990-2010, the production has increased significantly (Table 4.1).

Table 4.1. Average area harvested, yield and production of Soybeans in Nigeria for the past 51 years

Year	Area harvested (10^6 ha)	Yield (t ha^{-1})	Production (10^6 tons)
1961-1965	0.91	0.34	0.31
1966-1970	0.88	0.35	0.31
1971-1975	1.00	0.32	0.32
1976-1980	1.20	0.30	0.36
1981-1985	1.26	0.24	0.31
1986-1990	2.46	0.34	0.84
1991-1995	2.73	0.34	0.93
1996-2000	2.63	0.73	1.93
2001-2005	2.80	0.89	2.49
2006-2010	2.75	1.04	2.74
2011-2012	1.05	1.95	1.01

Source: http://faostat3.fao.org/home/index.html # HOME, 2012

Researches have been conducted on Soybeans in Mokwa, Zaria, Kano in Nigeria and in Chitedze in Malawi. The two locations in Nigeria represent two different agro-ecological zones of the moist savannah (Table 4.2). A total of 21 cultivars of Soybeans have been bred by the International Institute for Tropical Agriculture and were released in Africa. Most of these varieties were released in Nigeria. In terms of maturity, they were categorised as early, medium and late maturing. Among those released the grain yield of the early maturing ranged from 1.0 to 2 t ha^{-1}, while for the medium and late maturing, the yield ranged from 1.0 to 2.7 t ha^{-1} and 1.3 to 2.3 t ha^{-1} respectively (Tefera, 2011). Twenty-five medium maturing lines were developed from 1988 to 2006 for further utilization. The average maturity age ranged from 100 to120 days and the grain production ranged from 1.28 to 2.40 t ha^{-1}. Similarly, 20 promising lines of late maturing varieties (Table 4.3) have been developed between 1989 and 2005. These varieties matured between 107 to 123 days on the average under West African conditions. The late maturing varieties have been found to be of economic value for agro-ecologies with a long growth period and high rainfall depth (Tefera, 2011). This is one of reasons for selecting TGx 1448-2^E for cultivation in the current study.

Table 4.2. Some characteristics of Soybean breeding locations in Sub-Saharan Africa

Location	Coordinate	Elevation (m)	Rainfall (long term average mm)	Vegetation
Mokwa, Nigeria	6°5'N,9°48'E	308	900	Southern Guinea savannah
Zaria, Nigeria	11°11'N, 7°38'E	685	1100	Northern Guinea savannah
Kano, Nigeria	12°47'N, 9°2E	700	600	Sudan savannah
Chitedze, Malawi	15°55'S, 35°04'E	1146	892	Plateau

Source (Tefera, 2011). E = Early, M = Medium, L = Late

Table 4.3. The ranges of maturity and grain yields of late maturing promiscuous Soybean developed by the International Institute for Tropical
Agriculture (IITA) in the Guinea savannah of West Africa

Year	Line	Maturing date	Grain yield (t ha⁻¹)	Location
1989	TGx 297-10F	108-118(M)	2.0-2.1	Ghana
1989	TGx 297-192C	105-112(M)	1.4-1.8	Ghana
1989	TGx 306-036C	118-125(L)	1.5-2.0	Ghana, Nigeria
1989	TGx 888-49C	121-125(L)	1.4	Ghana
1989	TGx 923-2E	118-122(L)	1.5-2.0	Nigeria
1989	TGx 536-02D	100-110(M)	1.0-1.5	Ghana, Nigeria
1989	TGx 813-6D	105-110(M)	1.5-2.0	Ghana
1989	TGx 814-76D	110-117(L)	1.3-1.8	DR Congo
1989	TGx 849-294	97-103(E)	1.2-1.8	DR Congo
1991	TGx 849-313D	109-115(M)	1.4-1.8	Nigeria
1991	TGx 1019-2EN	98-106(E)	1.5-1.8	Nigeria
1991	TGx 1019-2EB	105-110(M)	1.5-2.0	Nigeria
1992	TGx 1440-1E	1115-120(L)	1.7-2.2	Nigeria (1992); Benin and Togo (1998)
1992	TGx 1448-2E	115-120(L)	1.7-2.3	Nigeria (1992); Benin and Togo (1998)
1998	TGx 1485-1D	85-95(E)	1.0-1.5	Nigeria (1992); Benin and Togo (1998)
1998	TGx 1740-2F	95-100(E)	1.0-1.5	Nigeria (1992); Benin and Togo (1998)
2005	TGx 1830-20E	190-93(E)	2.1	Ghana (2005)
2005	TGx 1904-5F	92-97(E)	-	Ghana (2005)
2004, 2009	TGx 1835-10E	90-95(E)	2.0	Uganda (2004), Nigeria (2009)
2007	TGx 1892-10E	121(L)	1.5	Ethiopia
2009	TGx 1904-6F	101-108(M)	2.5-2.7	Nigeria

Source (Tefera, 2011). E = Early, M = Medium, L = Late

4.5   Production practices in Nigeria

In Nigeria, the cultivation of the crop is often influenced by climate and soil characteristics. The crop grows well in the southern and northern Guinea savannah of the country where rainfall is more than 700 mm. However, short-duration varieties can thrive in the much drier Sudan savannah when sown early and with an even distribution of rainfall throughout the growing period. The time of sowing of the crop in Nigeria depends upon rainfall, soil temperature and day length. Soybean is a short-day plant and flowers in response to shortening days. It is grown on a wide range of soils with pH ranging from 4.5 to 8.5 (Dugje et al., 2009). Based on field experiences and observations, the suggested time of planting in the moist savannah/southern Guinea savannah is early June or early July while in the Northern Guinea savannah or Sudan savannah, mid June to early July is recommended. In the Sudan savannah, planting is done in the 1st to 2nd week of July. There are records of bacterial and fungal diseases that affect the crop in Nigeria and these include rust, caused by *Phakopsora pachyrhizi*; bacterial pustule, caused by *Xanthomonas axonopodis*; rogeye leaf spot caused by *Cercospora sojina*. Other recorded virus diseases of the crop are dwarf disease and yellow mosaic disease.

4.6   Prospects of Soybean production in Nigeria

Soybean is a relatively new crop in tropical Africa. It has long been thought that Soybean was not a suitable food crop for the region, because of the long cooking time needed and the unacceptable taste. However, the importance of the crop in tropical Africa has grown rapidly during the past decades. Especially Nigeria witnessed a rapid expansion in Soybean production in the smallholder farming sector in the savannah zone during the 1990s. The driving force for this expansion was the use of Soybean in the preparation of many traditional foods and the introduction of soya tofu, which rapidly became one of the most popular snacks in markets in the region and is widely used by the food processing industry. In some areas, the low world prices may depress opportunities for local producers to respond to increased local demand for Soybean. Soybean can play an increasingly important role in diversifying cereal-based farming systems in tropical Africa, especially in Nigeria.

Apart from being a source of residual nitrogen for subsequent cereal crops in crop rotations, the new multi-purpose cultivars of IITA provide the additional benefit that they help to reduce *Striga hermonthica* damage on Maize, Sorghum and Millet, thus representing a major opportunity to provide sustainable crop rotations for smallholder farmers. Soybean is cultivated in almost all parts of Nigeria with low agricultural inputs. The production of the crop in Nigeria had expanded because of its economic importance (Table 4.1). The rapid expansion of the poultry sector in the recent years has also increased demand for Soybean meal in Nigeria. As the farmers are getting more awareness about the crop, the production of the crop will increase in the nearest future. Based on the current trend, it is very likely that Soybean production will expand not only in Nigeria but also in many other tropical African countries in the future (Javaheri and Baudoin, 2001; Singh et al., 1987).

5 Crop yield models

5.1 Previous crop yield models

Efficient management of water under irrigated agriculture requires appropriate irrigation scheduling and accurate planting dates. The peasant farmers irrigate their crop based on experiences of the previous seasons. Similarly, other farmers irrigate their crops based on water availability without probing into the soil water status. In most cases, these practices have led to poor performance in terms of yield and wastage of water resources (Adekalu, 2004). Improvement in science and technology has increased human understanding of the interaction between soil, plant and atmosphere thereby making more precise irrigation scheduling possible, which has led to better performance of crops under excellent agronomic practices (Hillel et al., 1976). On a broad classification, two approaches are in use to develop models for solving problems in irrigation scheduling; these are statistical analysis and modelling. Statistical approach involves analysis of field data on crop water use and plant parameters. This approach has been used to determine the relationship between yield and irrigation depth, evapotranspiration, transpiration and soil moisture. The second approach involves modelling of physiological processes related to crop growth and its interaction with climate and soil. The two approaches described above have been used by irrigation scientists in order to predict yield in relation to water use and in modelling the water solute balance (Hanks and Hill, 1980; Cordova and Bras, 1981). Crop yield or growth models find application in irrigation planning and management and in environmental modelling. Unlike the statistical approach, modelling of crop growth is flexible, dynamic and allows transferability between experimental sites, crop varieties and years. Crop yield models are broadly categorized into phasic and non-phasic models. The phasic models divide the season into phenologic stages and evaluate the effect of water stress on each stage. However, the non-phasic models relate seasonal relative yield to seasonal relative total of one component of water use for instance evapotranspiration, transpiration and soil water storage (Adekalu and Okunade, 2008). The subsequent sections contain the various groups in which the crop yield models are categorized.

5.1.1 Group I models

The crop yield models in group I are based on the assumption that water is the only limiting factor influencing the growth of plants. The group I models were categorised into the following subgroups.

Input-output models

The first input and output model relates yields to number of irrigations and is expressed by Baird et al. (1987) as:

$$Y = F(N) \tag{1}$$

where:
Y   = actual crop yield (g m^{-2})
N   = number of irrigations

The second input-output model relates crop yield to irrigation depth and effective rainfall and is expressed as (Hanks and Hills, 1980):

$$Y = F\left(\sum R + \sum I\right) \tag{2}$$

where:
R   = rainfall (mm)
I   = irrigation depth (mm)

Potential deficit models

The potential deficit model relates yield to a form of potential evapotranspiration (ET_p) in relation to the total amount of water added in form of irrigation or rainfall. The deficit model appears in different forms. The active deficit model is expressed as (Penman, 1962):

$$Y = K \sum E_d + C \tag{3}$$

$$\sum E_d = \sum ET_p - (D - D_1) \tag{4}$$

where:

$$D = \left[ET_p - \left(\sum R + \sum I \right) + D_s \right] \tag{5}$$

where:
D_1 = limiting soil water deficit for optimum growth (mm)
ET_p = potential evapotranspiration (mm day^{-1})

The actual deficit model relates yield to actual evapotranspiration and is expressed as (Penman, 1962):

$$Y = K \sum ET_a + C \tag{6}$$

The simplest is the drought day (DD) model, which relates yield to the number of days the crop experienced drought that is when the soil moisture was less than 25% of the available water and is expressed as (Rickard and Fitzgerald, 1981):

$$Y = Y_m - gDD \tag{7}$$

where:
Y_m = maximum or potential crop yield (g m^{-2})
g   = average rate of yield loss (g m^{-2} year^{-1})
DD   = drought day

The De Wit model (De Wit, 1968) relates crop yield to transpiration and is expressed as:

$$Y = KT \tag{8}$$

where:
T = cumulative seasonal crop transpiration (mm year^{-1})
K = growth constant

The Hanks, H-1 model (Hanks et al., 1976) relates crop yield to transpiration and evapotranspiration purposely to allow transferability of models from one site season and cultivars to another and is expressed as:

$$Y = Y_m \times \left(\frac{T}{T_p} \right) \tag{9}$$

where:
Y   = actual crop yield (t ha^{-1})
Y_m = maximum yield (t ha^{-1})
T   = cumulative seasonal transpiration (mm year^{-1})
T_p = potential transpiration (mm year^{-1})

The Stewart (S-1) model is expressed as (Stewart et al., 1997):

$$Y = 1 - b \left(1 - \frac{\sum T_a}{\sum T_p} \right) \tag{10}$$

where:
b   = growth coefficient
ET_p = seasonal potential transpiration (mm year^{-1})
ET_a = seasonal actual evapotranspiration (mm year^{-1})

Despite the transferability of group I models, they are referred to as static models because they do not take into consideration the responses of crops to deficit water application at different growth stages. Therefore, they are used for large field yield reduction (Adekalu, 2004).

5.1.2 Group II models

The effects of deficit irrigation on crop yield can be precisely quantified by dividing the crop growth into stages. For this reason, the Hanks (H-2), Stewart (S-2), Hall-Butcher and Jensen models were developed from the earlier S-1 and H-2 models. The Stewart (S-2) and Jensen models relate crop yield to the actual evapotranspiration ratio. The Stewart (S-2) uses an additive model expressed as (Hanks et al., 1976):

$$Y_m = Y_m \left[1 - \sum_{i=1}^{n} K_{yi} \left(1 - \frac{ET_a}{ET_m} \right) \right]$$ (11)

where:

ET_i = actual evapotranspiration for stage i (mm day^{-1})
ET_p = potential evapotranspiration for stage i (mm day^{-1})
bi = growth stage weighing coefficient (-)

The modified Stewart et al. (1977) model was proposed for simulating the yield reduction caused by water deficit at different crop growth stages and is expressed as (Stegman et al., 1980b):

$$Y_a = Y_m \left[1 - \sum_{i=1}^{n} K_{yi} \left(1 - \frac{ET_a}{ET_m} \right) \right]$$ (12)

where:

K_y = Stewart's moisture stress yield reduction coefficient
Y_a = actual yield (t ha^{-1})
Y_m = maximum or potential yield (t ha^{-1})
ET_a = actual evapotranspiration (mm day^{-1})
ET_m = maximum evapotranspiration (mm day^{-1})

With little modification in the original expression, the Bras and Cordova function as reported by Igbadun (2007) is expressed as:

$$Y_a = Y_m \times \sum_{i=1}^{n} A \left(\frac{ET_a}{ET_m} \right)^i$$ (13)

where:

A_i = Bras and Corodova moisture stress sensitivity index for the growth stage 'i (no-unit)

The additive crop WP functions suggest that the water stress at a specific growth stage may not lead to total failure but may have significant impact on crop yield. The Hanks (H-2) model relates crop yield to the transpiration ratio and is expresses as:

$$Y = Y_m \times \prod_{i=1}^{n} \left(\sum \frac{T_i}{T_p} \right) \times \lambda i$$ (14)

where:

T_i = actual evapotranspiration for stage i (mm day^{-1})
T_p = cumulative potential transpiration (mm day^{-1})
λi = growth state weighing coefficient (mm day^{-1})

Hill et al. (1982) developed a multiplicative crop yield function based on root zone water deficit and growth stage and is expressed as:

$$Y = 100 \times \prod_{i=1}^{n} \left(\frac{T_a}{T_m} \right)^{\lambda i} \tag{15}$$

where:
Y   = relative yield (%)
N   = number of growth stages (usually between 4 and 6)
I   = growth stage (-)
Λ = fitted exponent (a calibrated value)
T_a = actual transpiration (mm day^{-1})
T_m = maximum transpiration (mm day^{-1})

The Hall-Butcher model relates crop yield to soil moisture ratio and is expressed as:

$$Y = Y_m \times \prod_{i=1}^{n} \left(\frac{W_{si}}{W_{sm}} \right) \times \lambda i \tag{16}$$

where:
W_{si} = soil water storage at the end of stage i (mm)
W_{sm} = available water storage (mm)

The Jensen model (Jensen, 1968) relates crop yield to evapotranspiration and is expressed as:

$$Y_a = Y_m \times \prod_{i=1}^{n} \left(\frac{ET_{ai}}{ET_{mi}} \right)^{\lambda i} \tag{17}$$

where:
ET_{ai} = actual crop evapotranspiration from moisture stressed treatment at growth stage 'i' (mm day^{-1})
ET_{mi} = crop evapotranspiration from non-stressed treatment at growth stage 'i' (mm day^{-1})
λ = Jensen's moisture stress sensitivity index (-)
N   = number of growth stages

The Minhas et al. (1974) function is expressed as:

$$Y_a = Y_m \times \prod_{i=1}^{n} \left[1.0 - \left(1.0 - \frac{ET_a}{ET_m} \right)^2 \right]^{\delta \times i} \tag{18}$$

where:

δ = Minhas's moisture stress sensitivity index (-)

ET_a = actual crop evapotranspiration from moisture stressed treatment at growth stage 'i' (mm day^{-1})

ET_m = evapotranspiration from non-stressed treatment at growth stage "i" (mm day^{-1})

Unlike the Stewart (S-2) model, Hanks (H-2), Hall-Butcher and Jensen models are *multiplicative,* in which the water stress at each growth stage is assigned a different weighing factor. Similarly, under a multiplicative model, the effect of water stress on crop yield depends on the growth stage. The multiplicative Crop Water Production Functions (CWPF) suggest that the effect of water stress on one or two crop growth stages may reduce the crop yield in a multiplicative manner. This implies that crops may experience a total failure if there is no evapotranspiration. These models are dynamic and useful in the evaluation of deficit irrigation at different phonologic stages of growth on the ultimate yield. A major challenge in the application of these models is that the potential yield of the crop under investigation must be known or given and that the growth stage coefficients must be obtained by calibration from the field data under different climatic and environmental conditions (Adekalu, 2004).

5.1.3   Group III models

The group III models were developed based on the link between plant physiology and transport of water and solute. In order to use this model, the daily net short-wave energy incident at crop height is used in calculating the PAR falling per square metre of crop. These values are combined with the simulated green LAI in order to compute the incident PAR of each hour at the midpoint of each leaf area by using the radiation interception model. The amount of CO_2 is fixed per layer per hour in the form of photosynthate. It is computed by using a photosynthesis equation in order to determine the CO_2 photosynthate value. Among the group III models are those of Zur and Jones (1981) and Ritche and Otter (1985) which are expressed as:

$$Y_m = CF \times PG \tag{19}$$

where:

CF   = carbon dioxide photosynthate value

PG   = photosynthate assimilation rate

Generally, expected maximum yield under a given climatic condition is obtained with these models. The maximum yield is later reduced by adverse physical factors such as water stress, aeration, limited fertilizer application and temperature. These models are useful in modelling environmental induced impacts such as insect infestation, canopy density and intercropping and effects of climate change on crop yield. However, these models are not adapted to irrigation scheduling because stress induced by water shortage is treated as a growth constraint alongside other factors, which are assigned values ranging from 0-1 depending on the ratio of the actual to potential evapotranspiration or transpiration. By using these models in quantifying the effects water stress on crop yield requires a lot of instrumentation, which in most cases is not available (Adekalu, 2004).

5.2   Computer based crop yield models: AquaCrop

5.2.1   Introduction

The AquaCrop model was developed from the previous Doorenbos and Kassam (1979) approach that is, an empirical function where relative evapotranspiration (ET) is very important in calculating yield (Steduto et al., 2007; Hsiao et al., 2007; Raes et al., 2009). The function is expressed as:

$$\left(1 - \frac{Y}{Y_x}\right) = K_y\left(1 - \frac{ET}{ET_x}\right) \tag{20}$$

where:
Y_x = maximum yield (t ha^{-1})
Y = actual yield (t ha^{-1})
ET_x = maximum evapotranspiration (mm)
ET = actual evapotranspiration (mm)

$(1-Y/Y_x)$ is relative yield decline due to water stress, the $(1-ET/ET_x)$ is relative decline in crop water use due to stress and K_y is proportionality factor between decline in the yield and relative reduction in evapotranspiration. The AquaCrop model was developed from the proportionality factor by separating actual evapotranspiration (ET) into soil evaporation (E) and crop transpiration (T$_r$). The partitioning of ET into Transpiration (T$_r$) and Evaporation (E) avoids the confounding effect of the non-productive consumptive use of water (E), which is important especially during incomplete ground cover or due to sparse planting of plants on the field. The AquaCrop model calculates the daily water balance and separates ET into its components, (T$_r$) and (E) based on the modification of Ritchie's approach (Ritchie, 1972) and this is a major advantage of the model. Furthermore, the AquaCrop model separates the final yield (Y) into biomass (B) and harvest index (HI):

$$Y = HI \times B \tag{21}$$

The separation of the final yield into the HI and biomass allows for the portioning of the functional relations as response to prevailing environmental conditions and their separation avoids the confounding effects of water stress on both biomass and HI. The changes described in the above equations lead to the formulation of the equation, which serves as the AquaCrop growth engine and is expressed as:

$$B = WP \times \sum T_r \tag{22}$$

where:
WP   = water productivity parameter (kg of the biomass m^{-2} and per mm of cumulated water transpired over the period of the time in which the biomass is produced
T_r = crop transpiration (mm)

The functional relationship between the components of the AquaCrop model is shown in Figure 5.1. The AquaCrop model has four main components namely the climate, crop, management and soil. The description of the components of the model is contained in the sections that follow.

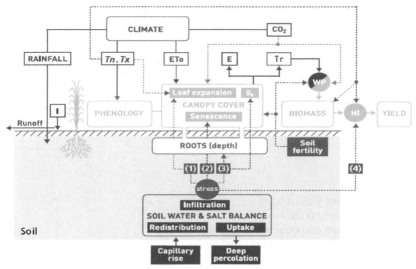

Figure 5.1. AquaCrop indicating the functional relationship between the components of the model in the soil-plant-atmosphere continuum and the parameters driving the phenology, canopy cover, transpiration, biomass production and final yield. Irrigation (I);Min air temperature (T$_n$); Max air temperature (T$_x$); Reference evapotranspiration (ET$_o$); Soil evaporation (E); Canopy transpiration (Tr); Stomatal conductance (gs); WP; HI; Atmospheric carbon dioxide (CO$_2$) concentration; (1), (2), (3), (4) represent different water stress response functions. Continuous lines indicate direct links between variables and processes. Dashed lines indicate feedbacks (Raes et al., 2012).

5.2.2 Description of the components

The atmosphere

The atmospheric environment of a crop under investigation is described in the climate component of the model. It requires five weather input variables and these are daily maximum and minimum air temperatures, daily rainfall, reference evapotranspiration (ET$_o$), which is daily evaporative demand of the atmosphere and the mean annual carbon dioxide concentration in the bulk atmosphere (Figure 5.1). The first four variables are obtainable at agrometeorological stations, while the CO$_2$ concentration uses the data recorded at Maunna Loa Observatory in Hawaii, USA and concentration for future years are entered by the use of the model. The reference evapotranspiration (ET$_o$) should be determined from daily SR, temperature, relative humidity and wind speed by using the model prescribed in FAO, Paper 56 (Allen et al., 1998) and is expressed as:

$$ET_o = \frac{0.408\Delta(R_n - G) + \gamma\left(\dfrac{900}{T+273}\right)u_2(e_s - e_a)}{\Delta + \gamma(1 + 0.3u_2)} \tag{23}$$

where:

ET_o	= reference evapotranspiration (mm d^{-1})
R_n	= net radiation at the crop surface (MJ m^{-2} d^{-1})
G	= soil heat flux density (MJ m^{-2} d^{-1})
γ	= psychometric constant (KPa $^{\circ}$C^{-1})
T	= mean of the monthly maximum and minimum air temperatures ($^{\circ}$C)
u_2	= wind speed at 2 m height (m s^{-1})
e_s	= saturated vapour pressure (KPa)
e_s	= actual vapour pressure (KPa)
e_s-e_a	= saturated vapour pressure deficit (KPa)
Δ	= slope vapour pressure curve (KPa $^{\circ}$C^{-1})

Paper 56 also described the methods for calculating reference evapotranspiration. Separate software called ET_o calculator, based on Paper 56, is used to estimate ET_o. 10-day or monthly data that can be processed by the AquaCrop model into daily values by using downscaling procedures. This is one advantage of the model that can be used to simulate and predict plant development with limited data. The AquaCrop model has a sub-routing to estimate effective rainfall from 10-day or monthly data by using two approaches. These are the USDA Soil Conservation Service method (Soil Conservation Service (SCS), 1993) or by setting effective rainfall as percentage of total rainfall. Temperature influences crop development and phenology, production of biomass and pollination in plants. Rainfall and ET_o are inputs for the water balance in the root zone soil and CO_2 concentration in the atmosphere enhances canopy growth rate and WP.

Soil

The soil components of the AquaCrop model are configured as horizons of variable depths. It allows up to five layers with different textural classifications in the soil and the values have to be specified by the use of the model. In the model, the soil hydraulic characteristics considered are: field capacity (FC), which is the upper limit of volumetric water holding capacity; permanent wilting point (PWP) taken as the lower limit of water holding capacity; drainage coefficient (τ), which is the ease at which water drains from the soil between saturation and FC, and hydraulic conductivity at saturation (K_{sat}). The textural classes in the USDA triangle (SCS, 1993) are include in the model, and can estimate the hydraulic characteristics according to textural class through pedotransfer functions (Saxton et al., 1986). In order to ensure high accuracy in the use of the model, the use need to specify the values that are applicable to a location. When the user specifies the soil data, estimates of Curve Number and readily evaporable water (REW) are generated automatically by the model. Within the root zone of the plants, the model performs a daily water balance by monitoring infiltration, runoff, internal drainage, root extraction in different depth layers, deep percolation, evaporation, transpiration and capillary rise. On daily basis, the model keeps records of both incoming and outgoing water fluxes and changes in the soil water content within the root zone. The model simulates water uptake by determining a root extraction term S as reported in the Reference Manual (Raes et al., 2012). The AquaCrop model uses the extent of green canopy cover to separate soil evaporation from transpiration. Soil E is taken to be basically proportional to the area of soil not covered by the canopy, but adjusted by using empirical equations for the effects of micro advection Raes et al. (2009). Soil evaporation follows the classical theory of bare-soil evaporation in which only Stage I (the energy limited phase) and Stage II (the declining phase limited by the transport of water to the soil surface and hydraulic properties of the soil) are considered as reported by Steduto et al. (2009a). The cumulative stage 1 evaporation, which is known as readily evaporable water (REW), is determined by using:

$$REW = 1000 \times (\theta_{FC} - \theta_{air\,dry}) \times Z_{e,surf} \qquad (24)$$

where:

FC = field capacity of the soil ($m^3\,m^{-3}$)

$\theta_{air\,dry}$ = soil water content when the soil is dry ($m^3\,m^{-3}$)

$Z_{e,surf}$ = thickness of the soil evaporating layer usually taken as 40 mm

When all the REW is removed, evaporation switches to stage 2, which is the falling rate stage. For both stages on bare soil, evaporation (mm) is determined by using the equation:

$$E = Kr \times E_x = K_r \times Kc_{e,wet} \times ET_o \qquad (25)$$

where:

$Kc_{e,wet}$ = evaporation coefficient for fully wet and unshaded soil surface (1.10 in Allen et al., 1998)

Kr = evaporation reduction coefficient

In order to account for the sharp reduction in hydraulic conductivity with reduction in soil water content, an exponential Equation 26 is used.

$$0 \le Kr = \frac{\exp^{f_K \times W_{rel}} - 1}{\exp^{f_K} - 1} \le 1 \qquad (26)$$

where:

f_K = decline factor

W_{rel} = relative water content of the soil

The rate of stage I evaporation when a soil is covered by a crop is given as:

$$E_x = (1 - CC^*) \times Kc_{e,wet} \times ET_o \qquad (27)$$

where:

E_x = evaporation from bare soil

CC* = adjusted fraction of soil surface adjusted for micro-advective effects.

At the late stage when CC reduces depending on the growth cycle or water stress, evaporation is given as:

$$E_x = (1 - CC^*) \times (1 - f_{CC} \times CC_{top}) \times Kc_{e,wet} \times ET_o \tag{28}$$

where:
f_{CC} = adjustment factor expressing the sheltering effect of dead canopy cover (≤ 1)
CC_{top} = canopy cover before the senescence

For soil covered by mulches, soil evaporation is adjusted by using the equation:

$$E_{x,adj} = E_x \left(1 - f_m \frac{\% \, fully \, mulched}{100} \right) \tag{29}$$

where:
f_m = adjustment factor for the effects of mulches on soil evaporation (ranges between 0.5 for plant materials and 1.0 for plastic mulches)

Adjustment is made for partial wetting under irrigated conditions by using:

$$E_{x,adj} = E_x \times f_w \tag{30}$$

where:
$E_{x,adj}$ = adjusted soil evaporation (mm)
f_w = fraction of the wetted soil surface

The AquaCrop model uses a function that is dependent on water content of the thin top soil layer for this purpose, in order to give a better reflection of E under conditions of low and high evaporative demand unlike the time dependent function used in other models for the stage 2 evaporation. Stage 2 evaporation is determined at a fraction of a day.

Crop

In the AquaCrop model, the crop system has five major components and associated dynamic responses (Figure 5.1): phenology, foliage canopy, rooting depth, biomass production, and harvestable yield. Crop response to water stress, can occur at any time along the crop phenology, through four major control links, which are stress coefficients (Ks) namely: reduction of canopy expansion rate (occurs during initial growth), closure of stomata (throughout the life cycle of the plant), acceleration of canopy senescence (especially during late growth), and changes in HI (after the commencement of reproductive growth). Extent of green canopy cover and duration represent the sources for transpiration, and the amount of water transpired is directly proportional to the amount of biomass produced through WP. Yield, which is the harvestable portion of the biomass, is determined from the product of biomass and HI. The model does not partition biomass into different components and organs because of the complexities and uncertainties in doing it. The interdependence between shoot and root in the model is not rigid and mostly indirect. The effects of water deficit reflect on canopy expansion and senescence. The rate of elongation of roots in the soil is linked to canopy via its growth and empirical function, which depends on effects of stress on the stomata.
The crop component has five sub-components, which are phenology, aerial canopy, rooting depth, biomass production and harvestable yield. The crop grows and

develops over its cycle by expanding its canopy and deepening its rooting system while progressing through its phenological stages. Responses of crop to water stress in the AquaCrop model occur through reduction of the canopy expansion rate, acceleration of senescence and closure of stomata, the WP parameter and the HI (Raes et al., 2009; Studeto et.al, 2009a).

Phenology and crop type

Phenology and crop development are cultivar specific and affected by temperature regimes. The AquaCrop model uses thermal time, growing degree days (GDD) as the default clock but runs on calendar only in day time steps. Users of the model are allowed to choose between the calendar day and GDD. There are three methods for determining GDD and details can be found in McMaster and Wilhelm (1997). In the model, GDD is determined by using method 2 bearing in mind that no modification is made for the minimum temperature when it falls below the base temperature. The model accommodates four major types of crops namely: fruit or grain crops; roots and tuber crops; leafy vegetable crops and forage, which can be harvested by cutting several times in a season. The key developmental stages considered for all the crops are: emergence, start of flowering (anthesis) or root/tuber initiation, maximum rooting depth, commencement of canopy senescence and physiological maturity. The extension of canopy with time depends on the determinacy of the crop and the users are allowed to vary it during calibration of the model. The cultivar under investigation or study needs to be evaluated by using the specified parameters of the generic crop stored in the crop file and necessary adjustments are made as appropriate.

5.2.3   Water productivity and above ground biomass

Biomass water productivity is very important to the operation of the AquaCrop model because its growth engine is driven by water availability (Figure 5.1). There are numerous experimental outputs to show that WP for many crops is conservative as reported in Steduto et al. (2009a). It has also been found to be conservative under water and salinity stress with low sensitivity to nutrient efficiency. In the model, the WP is normalized for climate and taken as a constant value for crops not minding the water stress except in extreme cases. Users can select from different categories of effects of soil fertility. Normalization of WP in the model is based on reference evapotranspiration and concentration of CO_2 in the atmosphere. Normalized water productivity (WP*) is determined by using the expression:

$$WP^* = \left[\frac{B}{\Sigma \left(\dfrac{T_r}{ET_o} \right)} \right]_{[CO_2]} \tag{31}$$

where:
B   = summation of biomass generated over the time intervals it is produced (g m^{-2})

The CO_2 outside the bracket indicates that the normalization is for a given year with specific mean annual CO_2 concentration. Due to variability in CO_2 concentration, it can be adjusted by using:

$$WP^*_{adj} = f_{CO_2} \times WP^* \tag{32}$$

$$f_{CO_2} = \frac{\left(C_{a,i} / C_{a,i}\right)}{1 + \left(C_{a,i} - C_{a,o}\right)\left[\left(1 - w\right)b_{Sted} + w\left(f_{sink}\right)b_{FACE}\right]} \tag{33}$$

where:

WP* = WP adjusted for CO_2
f_{CO2} = correction coefficient for CO_2
Ca,o = reference atmospheric CO_2 concentration (369 ppm)
Ca,i = atmospheric CO_2 concentration for year i (ppm)
b_{Sted} = 0.000138 (Raes et al., 2012)
b_{face} = 0.001165 (derived from FACE experiments)
w = weighing factor
f_{sink} = crop sink strength coefficient

The goal of normalization is to make the value of WP* specific for each crop and applicable to other locations and seasons and to simulate future climate scenarios. The AquaCrop model computes daily above ground biomass (B_i) by using the equation (Raes et al., 2012):

$$B_i = WP^* \left(\frac{T_{r_i}}{ET_{o_i}}\right) \tag{34}$$

where:

WP* = as defined previously
(Tr_i) = transpiration in a specific day
ET_{oi} = reference evapotranspiration from a specific day

A single value of WP* is used for the entire crop cycle. For crops where the proportion of lipids and protein in the harvestable yield is high, more energy is required per unit of dry weight produced after the grain or fruit begins to grow than before. In order to accommodate this, the AquaCrop model separates pre-anthesis and post-anthesis WP* by providing a fractional adjustment that reduces WP*. Irrespective of the daily T_r and the ET_o, biomass production can be affected by low temperatures beyond the level provided for by the thermal engine for crop development that is (GDD). Limitation imposed by low temperature and chemical composition of the harvestable organ is simulated with an adjustment factor that reduces WP* below normal values as a function of GDD. This adjustment is determined by using the equation:

$$m = Ks_b \times f_{WP} \times WP^* \times \left(\frac{T_r}{ET_o} \right) \tag{35}$$

where:

m   = adjusted daily above ground biomass production

Ks_b = adjustment factor for the effects of low temperature on production of biomass. Upper threshold is the minimum number of GDD required to achieve full conversion of Transpiration into biomass while the lower limit is the the point at which no Tr is converted into B

(f_{wp}) = adjustment factor to account for differences in chemical composition of the vegetative and harvestable organs

5.2.4 Responses to water stress

The timing, severity and duration of water deficit determine its impact on productivity and yield. In the AquaCrop model, four stress effects are identified. These are on leaf growth, stomata conductance, canopy senescence and HI. Except HI, the degrees of these effects are simulated by using an individual stress coefficient (k_s). It varies in value from 0 when there is full stress to 1 when the stress does not exist. For instance, k_s for water stress is a function of fractional depletion (p) of the total available water (TAW) and its value ranges between the upper and lower threshold for a specific crop. The evaporative demand of the atmosphere affects the rate of transpiration. In order to account for the effects of evaporative demand on leaf expansion, the lower and upper threshold of the stress response factor for stomata closure and leaf expansion of the day are adjusted relative to the ET_o, which is set at 5 mm day^{-1}.

Leaf expansion is the most sensitive of plant processes to water stress and stomata conductance and leaf senescence are less affected as reported in the AquaCrop Reference Manual (Raes et al., 2012). For stomata and senescence, the lower threshold is fixed at p = 1 that is (PWP) while that of leaf expansion is adjustable and should be reduced well below 1. There is no water stress coefficient (k_s) for HI in the AquaCrop model. The effects of water stress on HI are linked to the water stress coefficient for leaf growth and stomata conductance and indirectly to the water stress coefficient for senescence due to reduction in canopy duration.

Canopy

Canopy is an important component of the AquaCrop model. It determines the amount of water transpired through its expansion, aging, conductance and senescence. The amount of water transpired in turn determines the amount of biomass produced. Based on this, the foliage development in the model is expressed by using canopy cover (CC) not LAI. This feature distinguishes the AquaCrop model from other models. This simplifies the simulation of foliage development in the model as the CC can be estimated by an eye on the field. In addition, CC can be easily obtained from Remote sensing sources and imputed into the model for simulation. For the first half of the growth curve under nonstressed conditions, CC is expressed by using an exponential equation:

$$CC = CC_o e^{CGC \times t}$$ (36)

where:
CC   = canopy cover at time t or fraction of the ground cover
CC_o = initial canopy size at t = 0 in fraction
CGC   = canopy growth coefficient in fraction of per GDD or per day, which is a
              constant for a crop under optimal conditions and can be influenced by
              stresses. CC_o is proportional to plant density and average initial canopy size
              per seedling (cc_o) which is used to account of the variation in plant density

Based on data obtained from the field on the heterogeneity of germination, foliage canopy at 90% germination is taken as cc_o and is taken as representative for the population of plants on a field. For most crops, the CC_o is conservative for instance for Maize (Hsiao et al., 2009). For other crops, the use of the AquaCrop model needs to input the plant density and initial canopy cover will be automatically generated. At the second half of the CC curve, canopy growth follows an exponential decay and this is due to effects of shading. At this stage, CC is expressed as:

$$CC = CC_x - (CC_x - CC_o) \times e^{-CGC \times t}$$ (37)

where:
CC   = is the maximum CC under optimal conditions and the default values are
              provided in the AquaCrop model

Since maximum CC (CC_x) is determined by factors such as plant density, management criteria, default values in the AquaCrop model can be adjusted to the actual field conditions during simulation. Figure 5.2 shows the exponential growth and decay stages.

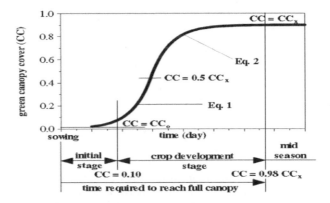

Figure 5.2. Schematic representation of canopy development during the exponential growth and the exponential decay stages (Raes et al., 2012)

Since foliage growth is very sensitive to water stress, development stages of the plant can be influenced by water stress, which results from fractional depletion of the total available water. The effects of water stress are estimated by using the equation:

$$CGC_{adj} = k_{s_{exp}} CGC \tag{38}$$

where:
CGC$_{adj}$ = adjusted canopy growth coefficient (per day)
ks$_{exp}$ = water stress coefficient for canopy expansion and ranges between 0 and 1

Water stress may prevent the CC$_x$ from being reached thereby resulting in smaller CC$_x$. This is often the case in determinant crops whose canopy growth is permitted to the middle of flowering in the AquaCrop model. If water stress is severe enough, senescence in plants can occur even at the middle of the development stage.

As the crop tends towards physiological maturity, senescence causes decline in CC. In the model decline in green CC is expressed by:

$$CC = CC_x \left[1 - 0.05 \left(\exp^{\frac{CDC}{CC_x}t} - 1 \right) \right] \tag{39}$$

where:
CDC = canopy decline coefficient expressed in fraction reduction per GDD or per day
T   = t is time since the beginning of senescence of canopy

Typical shapes of canopy decline are shown in Figure 5.3. In the model , canopy decline occurs later than leaf senescence because senescence starts with old leaves located under the canopy and contributes less to T$_r$ and biomass accumulation.

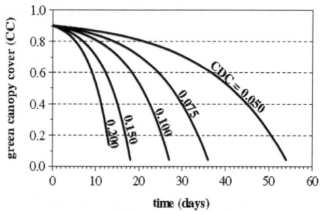

Figure 5.3. Reduction in green canopy cover (CC) during senescence for various canopy decline coefficients (CDC). All curves have initial green canopy cover (CC$_o$) at 0.9 and starting time at 0 (Raes et al., 2012.)

In the AquaCrop model, canopy decline is triggered when T$_r$ and photosynthesis are declining towards maturity. The progression of CC over a full crop cycle is shown in Figure 5.4.

Senescence of a canopy can be speeded up at any time in the life cycle of a plant if the stress is strong enough. The effects of water stress are simulated by adjusting the

CDC by using the equation:

$$CDC_{adj} = (1 - Ks_{sen}^8) \times CDC \qquad (40)$$

where:
CDC$_{adj}$ = adjusted canopy decline coefficient (per day)
(Ks_{sen}) = the water stress coefficient for the acceleration of senescence

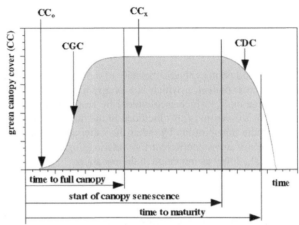

Figure 5.4. Variation of green CC from emergence until physiological maturity under non-stress conditions (Raes et al., 2012)

Transpiration

Where there is no stress induced stomata closure, transpiration is proportional to CC with adjustment for interrow micro advection and sheltering effects by partial canopy. This indicates that evaporation from the soil is proportional to (1 - CC). The effects necessitated the adjustment of CC in order to determine T$_r$. Where there is no water stress, transpiration is determined by using the equation:

$$T_r = CC^* \times K_{c_{tr,x}} \times ET_o \qquad (41)$$

where:
T$_r$ = transpiration (mm)
CC* = adjusted canopy cover
$K_{c_{tr,x}}$ = coefficient for maximum crop transpiration
ET$_o$ = as previously defined

After reaching the CC$_x$ and shortly before senescence, canopies undergo slow aging and progressive reduction in T$_r$ and photosynthetic capacity of the plant. This effect is simulated by introducing a factor called ageing coefficient (f_{age}), which reduces K_{cb_x} by a constant slight fraction of about 0.3% per day. When senescence fully begins, T$_r$ and photosynthetic ability of the plant is reduced. Therefore, T$_r$ is reduced by a specific reduction coefficient called (f_{sen}), which declines from unity (1) at the commencement of senescence to 0 when no green CC remains (CC = 0). These are illustrated by using the equations:

$$K_{c_{tr,sen}} = f_{sen} \times K_{c_{tr,adj}} \tag{42}$$

$$f_{sen} = \left(\frac{CC}{CC_x}\right)^a \tag{43}$$

$$T_{r_x} = CC^* \times K_{c_{tr,sen}} \times ET_o \tag{44}$$

Reduction in T_r and photosynthetic capacity of a plant can be increased (a > 1) or decreased (a < 1) by an exponent a, which is a program parameter in the model. The effect of water shortage on T_r, which is triggered by stomata closure or waterlogging that triggers anaerobiosis on crop, is also included in the model. This effect is simulated by multiplying maximum transpiration by either the water stress coefficient for stomata closure (ks_{sto}) or the water stress coefficient for waterlogging (Ks_{aer}) to obtain the actual transpiration (Raes et al., 2009) as indicated in the equation:

$$T_r = K_s \times T_{r_x} \tag{45}$$

The extraction out of the root zone of the water lost by T_r is based on a sink term that is the rate of extraction per unit of soil volume (m^3 m^{-3} d^{-1}). In order to obtain the sink term at specific soil depth, the maximum extraction term is modulated by a water stress factor at that depth (K_{si}) by using the equation:

$$S_i = K_{s_i} \times S_{x,i} \tag{46}$$

where:
S_i = sink term at specific depth
K_{si} = water stress coefficient for stomata closure or anaerobiosis

The cumulative extraction rate is matched to T_r by integrating Equation 46 over the entire rooting depth, starting from the top of the soil profile and ending when the sum (1000Si Δz_i) is equal to Tr or when the bottom of the root zone is reached:

$$Tr = \int_{i=top}^{i=bottom} 1000 \left(Ks_i \times S_{x,i}\right)\Delta z_i \qquad \leq Ks \times T_{r_x} \tag{47}$$

5.2.5   Root extension and water extraction

In the AquaCrop model , the root system is simulated through effective rooting depth and its water extraction pattern. A pattern different from the default values 40, 30, 20 and 10% can be adopted by the user. Capacity for water extraction is modulated by using water extraction S_i (Raes et al., 2009). The elongation of the root until it reaches the maximum depth along the crop cycle is determined by using the empirical equation:

$$Z = Z_{ini} + (Z_x - Z_{ini}) N \sqrt{\frac{\left(t - \frac{t_o}{2}\right)}{t_x - \frac{t_o}{2}}} \qquad (48)$$

where:

Z	= effective rooting depth at time t (days) after plating
Z_{ini}	= depth at sowing
Z_x	= maximum effective rooting depth
t_o	= time after planting to effective (85-90%) emergence of the crop (GDD or day)
t_x	= time after planting when Z_x is reached
N	= shape factor of the function

Root deepening or rate of elongation should be at its maximum and is expected to be attained near the end of the crop cycle. However, the presence of a restrictive soil layer may reduce or stop root penetration. Root elongation is programmed to reduce in the model when the depletion in the root zone exceeds the threshold for stomata closure (Steduto et., 2009a). The reduction is determined by the magnitude of the water stress coefficient for stomata closure and a shape factor (-ve) by using the equation:

$$dZ_{adj} = dZ \frac{e^{K_{sto} f_{shape}} - 1}{e^{f_{shape}} - 1} \qquad (49)$$

where:

dz	= rate of deepening of roots
f_{shape}	= default value of (-6)
K_{sto}	= effect of water stress in the root zone

Figure 5.5 shows the generalized schematic representation of development of rooting depth with time.

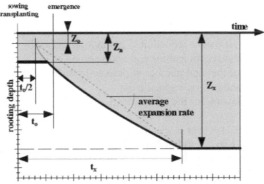

Figure 5.5. Schematic representation of rooting depth along the crop growth cycle from sowing until maximum (shaded area) effective rooting depth (Raes et al., 2012)

5.2.6   Responses of harvest index (HI) to water stress

HI is the proportion of the harvestable part of the biomass. The value that is commonly obtained at maturity in absence of water stress is taken as reference for a specific case.

In the AquaCrop model, HI increases with time after a lag phase until physiological maturity. Effects of water stress on HI depend on timing and severity. If the vegetative growth is still feasible, HI is enhanced by limiting foliage growth. The rate of increase in HI speeds up if the threshold for canopy expansion is reached and the water stress coefficient for stomata closure falls below 1 as water depletes by using the equation:

$$\frac{dHI}{dt} = \left(1 + \frac{(1 - Ks_{exp,i})}{a}\right) \times \left(\frac{dHI}{dt}\right)_o \tag{50}$$

where:
dH/dt = rate of increase of HI for non stressed conditions
a = adjustable crop parameter

This equation is relevant as long as vegetative growth occurs even if CC is already complete. Further imposition of water stress inhibits stomata opening and the cumulative reduction in supply of assimilates, which slows down the increase in HI is described by using the equation:

$$\frac{dHI}{dt} = \sqrt[10]{Ks_{sto}}\left(1 + \frac{(1 - Ks_{sto,i})}{b}\right) \times \left(\frac{dHI}{dt}\right)_o \tag{51}$$

where:
b = adjustable crop parameter whose negative effects on HI is increase as b is adjusted downward and reduces if b is adjusted upward

In situations where water stress is strong enough to inhibit pollination, the effects on HI are determined by using the equation:

$$HI_{o,adj} = \left[\sum Ks_{pol} \times a \times F\right] \times HI_o \leq HI_o \tag{52}$$

where:
$HI_{o,\ adj}$ = adjusted HI for the reduction in pollination caused by water stress
K_{spol} = water stress coefficient for pollination on a given day
F = fraction of the total number of potentially successful flowers that are going through anthesis on a specific day
a = factor allowing for the effects of excessive sinks (potential fruits)

It is a constant (> 1) for a specific crop. The value depends on excessive sinks that a crop produces. Fereres and Soriano (2007) suggested that HI can be enhanced if water stress occurs before anthesis and could be associated with reduction in pre-anthesis biomass. Adjustment of HI is computed by using the equation:

$$R_{ange}(B_{rel}) = \left(\frac{\ln(\Delta HI_{x,ante})}{5.62}\right) \leq 1 \tag{53}$$

where:
$\Delta HI_{x,\ ante}$ = maximum allowable increase in HI due to pre-anthesis stress (%)
B_{rel} = ratio of biomass to potential biomass (B_x-biomass at optimal conditions)

The model adjusts the HI on a daily basis during simulation for the effects of each of the stated stresses when required. The final HI is not allowed to go beyond the HI_o by a specific percentage specified during calibration or entered by the user.

Management

In the AquaCrop model, the management component has two main categories and these are field management and water management. The field management offers the user an option to choose from or define the: (i) fertility level the crop is exposed to during the crop cycle; (ii) field management practices to reduce evaporation such as mulching or controlled surface runoff e.g. use of soil bund and a small dyke; (iii) time of cutting the forage crop. The fertility level in the model ranges from non-limiting to poor with attendant accelerated reduction in normalized water productivity, canopy growth coefficient and maximum canopy cover as soil fertility decreases. The AquaCrop model does not compute the nutrient balance, but only offers the semiquantitative options, which can be used to access the effects of fertility levels on biomass production and final yield. Under the field management practices, height of the soil bund needs to be stated for the purpose of simulation.

Two options are available in the model under water management and these are: (i) rainfed agriculture and (ii) irrigation. The methods of application need to be selected from options that are available, which include drip, sprinkler or surface. The user defines the schedules by specifying the time and depths of application or allows the model to to automatically generate the schedule based on time intervals and depth for each application or fixed allowable water depletion.

5.2.7 Criterion for the model evaluation

The ability of a model to reproduce observed data is important in order to evaluate its performance. Several statistical indicators are available for achieving this and each has its own strength and weakness, which necessitates the use of several statistical indicators in order to evaluate the performance of a model (Retta et al., 1996). The following statistical indicators are used in the AquaCrop (Version 4.0) model to compare field observed and the simulated data (Raes et al., 2012).

Coefficient of determination (r^2)

The coefficient of determination r^2 shows the proportion of the variance in measured data explained by a model. Its values range between 0 and 1. Values close to 1 indicate a good agreement between measured and observed data. r^2 is expressed as:

$$r^2 = \left[\frac{\sum (O_i - \overline{O})(P_i - \overline{P})}{\sqrt{(O_i - \overline{O})} \times \sum (P_i - \overline{P})} \right] \tag{54}$$

where:
O_i = observed data

$\overline{O}$ = mean of observed data

P_i = predicted data

$\overline{P}$ = average of predicted data

Root Mean Square Error (RMSE)

RMSE is a measure for the average magnitude of the difference between predictions and observations. It ranges from 0 to positive infinity. The former indicating good and the latter poor model performance. RMSE does not differentiate between under and over estimation of values. It is expressed as:

$$RMSE = \sqrt{\frac{\sum(P_i - O_i)}{n}}$$ (55)

where:
n   = number of observations

Normalized Root Mean Square Error (NRMSE)

The NRMSE is expressed as percentage and gives an indication of the relative difference between model results and observations by the equation:

$$NRMSE = \frac{1}{\overline{O}}\sqrt{\frac{\sum(P_i - O_i)^2}{n}} \times 100$$ (56)

   A simulation is considered excellent if NRMSE is smaller than 10%, good if between 10 and 20%, fair if between 20 and 30% and poor if larger than 30% as reported in the AquaCrop Reference Manual (Raes et al., 2012).

Nash-Sutcliffe model efficiency coefficient (EF)

The Nash-Sutcliffe Efficiency coefficient (EF) indicates how well the plot of observed versus simulated data fits the 1:1 line. EF ranges between $-\infty$ and 1.0 (1 inclusive) with EF = 1 being the optimal value. An EF of 1 indicates a perfect match between the model and the observations, an EF of 0 means that the model predictions are as accurate as the average of the observed data and a negative EF occurs when the mean of the observations is a better prediction than the model (Moriasi et al., 2007). The Nash-Sutcliffe model efficiency coefficient is expressed as:

$$EF = 1 - \frac{\sum(P_i - O_i)^2}{\sum(O_i - \overline{O})^2}$$ (57)

Willmott's index of agreement (d)

The index of agreement is a measure of the closeness between the observed and the predicted data. It represents the ratio between the mean square error and the "potential error", which is defined as the sum of the squared absolute values of the distances from the predicted values to the mean observed value and distances from the observed values to the mean observed value as reported in the AquaCrop Reference manual (Raes et al., 2012). It ranges between 0 and 1, with 1 indicating a perfect agreement between the predicted and observed data and 0 indicating no agreement. It is expressed as:

$$d = 1 - \frac{\Sigma (P_i - O_i)^2}{\Sigma \left(\left| P_i - \overline{O} \right| + \left| O_i - \overline{O} \right| \right)^2} \tag{58}$$

However, d is still not very sensitive to systematic over or underestimation (Krause et al., 2005).

5.3   Radiation use efficiency (RUE)

When light falls on a plant, the pigments absorb the radiant energy thereby resulting in establishment of new energy levels in the molecules. The higher energy states are later used to assimilate carbon oxide (CO_2) and also synthesise plant constituents. The fraction of the radiation that is intercepted by a crop is also a function of the LAI and the extinction coefficient of the plant (λ). This is expressed in the Beers and Lambert law:

$$PAR_b = PAR_a \left[\left(1 - \exp(\lambda \times LAI) \right) \right] \tag{59}$$

where:
PAR_b = photosynthetically active radiation below the canopy (μmol m^{-2} s^{-1})
PAR_a = photosynthetically active radiation above the canopy (μmol m^{-2} s^{-1})

LAI is the ratio of the area of the green leaf to the area of the soil covered by it and is expressed as (m^2 m^{-2}). The LAI of crop varies from one area to the other and on management. Numerous values of λ have been reported in literature. For Soybeans (*Glyxine max (L.) Merrill*), Cowpea (*Vigna unguiculata (L.) Walp.*) and Muungbean (*V. Radiata (L.) Wilczek*) it ranged from 0.4 to 0.8 (Muchow, 1985) and a value of 0.3 has been reported for pigeon pea (*Cajanus caja (L.) Millsp.*). Values of 0.4 to 0.61 have been reported for chickpea by Hughes et al. (1987), of 0.33 to 0.49 for pea by Heath and Hebblethwaite (1985), which is within the range of values reported by Thomson and Siddique (1997) for many grain legumes. A λ of 0.93 was reported for Cowpea (Varlet-Grancher and Bonhommme in Jeuffroy and Ney, 1997).

Water and nutrient availability in soils play important roles in the development and growth of plants. Moisture deficit, especially during critical growth stages of plants, may hinder them from attaining their highest potential in terms of CC and yield. The amount of light available to plant canopy determines its yield and provides important information about the physiological processes and impacts of microclimate on the plant (Singer et al., 2011). Under field conditions and in the absence of water stress, growth of crops depends on the ability of the canopy to intercept incoming radiation. Several analyses of the increase in mass of crops in response to the amount of available light were done by De Wit (1959) and Loomis et al. (1963). Studies show that light levels have great effects on plant growth and considerable attention has been given to establish a scientific relationship between crop growth and the light level over the years in many parts of the world (Monteith, 1977; Monteith, 1994, Sinclair and Muchow, 1999). In the recent times, emphasis has been placed on the modelling of plant growth based on radiation capture by the canopy rather than the localised leaf area expressed by the assimilation model, which has been overly criticised (Sinclair and Muchow, 1999). PAR is the radiation in the 400 to 700 nm waveband. Interception of PAR by a crop can be in terms of the TIPAR or the fraction of this radiation, which is intercepted by the canopy. PAR is very important in modelling photosynthesis of either a single plant or plant communities (Alados et al., 1966). The total incident PAR depends on location, time of the year, and crop phenology, whereas the fIPAR is a function of LAI and λ (Monteith, 1965; Plenet et al., 2000). The value of λ is determined by the canopy structure, species and planting pattern. Values range from 0.3 to 1.3 (Saeki, 1960; Zarea, 2005). A λ less than 1 refers to non-horizontal or clumped leaf distributions and a value greater than 1 refers to horizontal or regular leaf distributions (Jones, 1992).

RUE is the biomass produced per unit of the total SR or PAR that is intercepted by the canopy and is a key determinant of the yield of a crop (Monteith, 1977; Stockle and Kemanian, 2009; Soltani and Sinclair, 2012). Under a non-stressed condition, the rate of biomass accumulation in a crop has been found to be a linear function of the TIPAR and the efficiency of the use of that SR (Monteith, 1975; Monteith, 1977; Kiniry et al., 1989). The slope of the relationship gives the rate at which the intercepted radiation is being converted to biomass in plant communities. Therefore, in order to model plant growth under water stress or non-stressed conditions, there is a need to adequately understand and describe LAI, λ for PAR and RUE.

Many papers summarised the RUE obtained under a wide range of environments (Gallagher and Biscoe, 1978; Gosse et al., 1986; Kiniry et al., 1989). Gosse et al. (1986) concluded that important differences exist in RUE between species and cultivars. Thomson and Siddique (1977) reported a RUE ranging from 0.41-0.99 g MJ^{-1} of IPAR for various legumes. Singer et al. (2011) reported a RUE of 1.46 g MJ^{-1} of IPAR for Soybeans. RUE ranging from 1.32 to 2.52 g MJ^{-1} of IPAR from six studies were reported for Soybeans by literature review of Sinclair and Muchow (1999) and 1.20 gMJ^{-1} IPAR was reported by Nakeseko et al. (1983). Leadley et al. (1990) reported a maximum RUE of 0.86 gMJ^{-1} of intercepted SR in Raleigh, USA. Rochette et al. (1995) reported a RUE of 2.04 g MJ^{-1} of IPAR in Canada, while Daughtry et al. (1992) reported a maximum RUE of 2.34 gMJ^{-1} of absorbed PAR for Soybean in Beltsville. RUE ranging from 0.66 to 1.15 g MJ^{-1} of the intercepted SR was reported by Sinclair and Shiraiwa (1993) in Japan. RUE of 1.08 g MJ^{-1} and 1.89 g MJ^{-1} of absorbed PAR was reported for Soybean grown under saline and non-saline conditions respectively (Wang et al., 2001). Schoffel and Volpe (2001) reported a mean value of 1.23 g MJ^{-1} of PAR for different cultivars of Soybean in Southeast Brazil. Confalone and Dujmovich (1999) recorded a RUE ranging from 1.37 g MJ^{-1} under irrigated conditions to 1.92 g MJ^{-1} of PAR under rainfed conditions in Argentina. Santos et al. (2003) reported RUE ranging from 2.28 to 2.53 g MJ^{-1} before and after flowering while Schoffel and Volpe (2001) found averages of 1.02 and 1.40 g MJ^{-1} at vegetative and reproductive sub-periods under water stressed conditions for many cultivars. Similarly, Confalone et al. (1998) reported 1.73 to 1.86 g MJ^{-1} of PAR between vegetative and flowering stages of Soybean.

There is no specific value for RUE of a certain crop. It varies based on species, environment, CO_2 assimilation rate and water stress during the growth stages (Muchow, 1985). Other factors that influence RUE include the radiation environment [vapour pressure deficit (VPD)], air temperature, water stress, growth stage of a crop and location (Sinclair and Muchow, 1999). Latitude and LAI have minor roles in explaining variations in RUE of crops. Radiations that reach plant communities are either intercepted or absorbed. (Sinclair and Muchow, 1999) reported that RUE in terms of the total SR is about 0.5 times the RUE of the IPAR and 0.425 times the RUE based on absorbed PAR.

According to Kiniry et al. (1999), increased understanding of factors that control the production of biomass is required to accurately simulate growth and quantify productivity in different environments. Soybean is cultivated in Nigeria and other African countries under rainfed conditions. It is an economic crop and offers many benefits. However, field data on λ and RUE of Soybeans are either scarce or non-existing in Ile-Ife. There is limited or no information on the relationship between yield, IPAR and RUE with water productivity of Soybeans.

5.4 Concept of irrigation water use efficiency and water productivity (WP)

Following the extensive work in 1940 by Israelsen, irrigation efficiency was described as the ratio of irrigation water consumed by crops of an irrigation scheme to the amount of water diverted from a river or other natural channel (Israelsen, 1950). For more than 40 years, this basic approach to irrigation water accounting remained unchanged. The quantity of water used in controlling salinity should also be considered as beneficial use in order to sustain irrigation. Irrigation efficiency was therefore defined as the ratio of consumptive use of crops and water required for leaching on a steady state basis to the volume of water diverted, stored, or pumped specifically for irrigation (Jensen, 1967). This definition made the water balance calculation complex. The results of a joint effort of the International Commission on Irrigation and Drainage (ICID), the University of Agriculture, Wageningen, and the International Institute for Land Reclamation and Improvement (ILRI), Wageningen were published in Bos and Nugteren (1974, 1982). More light was thrown on the definition of irrigation efficiency. Distribution efficiency was defined as the ratio of the volume of water furnished to the fields to the volume of the water delivered to the distribution system. Field application efficiency was also defined as the ratio of the volume of irrigation water needed, and made available for evapotranspiration by crops to avoid undesirable water stress in the plants throughout the entire growing cycle to the volume of water furnished to the fields.

CWP is defined as the ratio of the net benefits from crop, forestry, fishery, livestock and mixed agricultural systems to the amount of water required to produce those benefits. It can also be described as the value added to water in a given circumstance (Molden and Sakthivadivel, 1999; Tuong et al., 2000). In its broadest sense it can be described as producing more food, income, livelihoods and ecological benefits at less social and environmental cost per unit of water used. Physical water productivity is the ratio of the mass of agricultural output to the amount of water used. When WP of a specific crop is measured, it is referred to as CWP. CWP is a key technical term in the scientific evaluation of deficit irrigation (DI). The economic water productivity or (WP$_{economic}$) is defined as the value derived per unit of water used and the unit can be expressed in terms of any currency for instance (₦ m^{-3}) (Kadigi et al., 2004). It has become an issue of interest and evaluation to agronomist and irrigation engineers for about a century (Briggs and Shantz, 1916; De Wit, 1958; Hanks et al., 1969; Hanks, 1974; Angus et al., 1980; Sinclair et al., 1984; Howell et al., 1990; Musick et al., 1994; Angus and van Herwaarden, 2001. In literature, crop water productivity is also referred to as *water use efficiency (WUE)* (French and Schultz, 1984a; French and Schultz, 1984b, Zhang and Owie, 1999; Howell, 2001). CWP is a mean of assessing the effects of water supply on yields of crops and it aids management decisions on effective and sustainable planning of irrigation schemes in areas where water is very scarce (Augus and van Herwaarden, 2001). According to Kijne et al. (2003), Zwart and Bastiaanssen (2004) and Igbadun et al. (2006) CWP is expressed as:

$$CWP = \frac{Y}{SWU} \tag{60}$$

where:
CWP = crop water productivity (kg ha^{-1} mm^{-1})
Y = marketable crop yield (kg ha^{-1})
SWU = seasonal evapotranspiration (mm)

CWP is also expressed in terms of irrigation water applied (SWA) and is given as:

$$IWP = \frac{Crop\ yield\ (kg)}{SWA(m^3)} \tag{61}$$

where:
SWA = seasonal irrigation water applied (mm year^{-1})

Crop water productivity is also expressed in terms of the available price of the yield of crops in the market and is expressed as:

$$CWP = \frac{p \times crop\,yield}{SWA(m^3)} \tag{62}$$

where:
p = market price per kg of the crop in question

5.5   Simple model for carbon assimilation by plants

Carbon assimilation by plants can be described as the chemical transformation of carbon oxide and water to carbohydrate and oxygen within the leaves of plants. Energy in the form of light, mostly from the sun, is required in order for this process to proceed. The CO_2 comes from the environment (atmosphere) and must diffuse into the messophyll cell to be fixed. The amount of water used during photosynthesis is very small, but a large amount of water is required during the uptake of CO_2. Therefore, carbon assimilation can be limited by two factors namely: light and water. This led to the use of either light based or water based models in simulating dry matter production in plants.

5.5.1   Light based model

Monteith (1977) observed that when measured biomass accumulation by plants is plotted against accumulated SR that is intercepted by the plant community, in almost all cases the result was a straight line. This led to the suggestion of the model that the accumulated biomass (P) is proportional to the fraction of the intercepted radiation (f), flux density of the incident radiation intercepted by the canopy (S) and the conversion efficiency of the canopy (e). Conversion efficiency is quite conservative for a specific specie and in the range of 0.01 to 0.03 mol. CO_2 (mol photons)$^{-1}$ (Campbell and Norman, 1998). Mathematically this it is expressed as:

$$P = e \times f \times S \tag{63}$$

Flux density of the incident radiation is the only environmental factor, while fraction of the incident radiation intercepted by the canopy and conversion efficiency are determined by crop physiology and management. Because only visible wavelengths are effective in photosynthesis, PAR has been used to estimate canopy assimilation rather than SR. The effects of water stresses can be quantified in terms of reduction in conversion efficiency and fraction of the intercepted radiation.

5.5.2 Water based model

When water is limited, a water-based model is applicable and is expressed as:

$$A = \frac{kT}{D} \tag{64}$$

where:
A   = dry matter assimilation (kg ha^{-1})
k   = constant for a specific specie and atmospheric CO_2 level
D   = atmospheric vapour deficit
T   = transpiration (mm)

Potential Transpiration (T_p) is determined as the product of potential evapotranspiration and fraction of the radiation (f) intercepted by a canopy along its life cycle and is expressed as:

$$T_p = f \times ET_x \tag{65}$$

where:
ET_x = potential evapotranspiration (mm)

Similarly, potential evaporation of water from soil is computed from:

$$E_p = (1 - f) \times ET_x \tag{66}$$

where f and ET_x are as defined previously.

6   Research methodology

6.1   Study area

The experiments were conducted at theTeaching and Research Farms of Obafemi Awolowo University (TRFOAU), Ile-Ife located at latitude 7^0 28' 0"N and longitude 4^0 34' 0"E, 271 m+MSL during the rainy and dry seasons in the period 2011 till 2014. The TRFOAU is located at about 7 km away from the academic and administrative sections of the university and it covers an area of 1,400 ha (Figure 6.1). The land at TRFOAU is used mainly for crop cultivation and as a demonstration farm for teaching of students and conduct of researches. Ile-Ife is in the sub-humid (SH) agro climatic zone of Nigeria and is the major agrarian community and economic base of Osun state, Nigeria. It is located within Ogun-Osun River Basin.

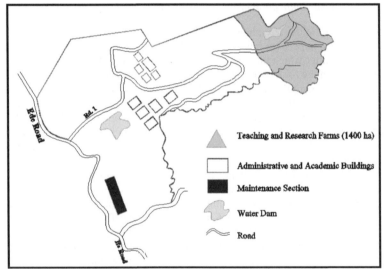

Figure 6.1. Location of the experimental fields at the Teaching and Research Farms of Obafemi Awolowo University, Ile-Ife

6.2   Experimental treatments, field lay out, cultivation and measurements in the rainy seasons

The experimental treatments, field lay out, cultivation practices and measurements made are presented in the sections that follow.

6.2.1   Experimental treatments

The experimental treatments consist of six water conservation practices and the conventional practice. They are as follows:
- *Tied ridges (TR)*. The ridges were constructed manually and tied together at intervals of 1.5 m along the rows by using heaps of soil and with a local iron hoe (Figure 6.2). This was done because of the nature of rainfall in the area;

- *Mulching (ML).* The plant materials (*Panicum maximum*) were spread on the ridges and ground below the plant and covered 75% of the soil surface in each plot. The total mulch application rate was 2.16 kg m^{-2} or 21.6 t ha^{-1} per treatment. The mulches were applied after the establishment and three times before the maturity of the crop;
- *Bunding (BD).* A *h*eap of soil (side bund) of 30 cm high was constructed manually around each plot to trap and concentrate the rainfall within the plot at the root zone of the plant. The bunds were repaired as occasion demanded;
- *Tied ridges plus mulching (TRML).* A combination of tied ridges and the use of mulches;
- *Tied ridges plus bunding (TRBD).* A combination of tied ridges and bunds;
- *Bunding plus mulching (MLBD).* A combination of side bunds and mulches;
- *Non conservation measures or Conventional practice (NC).* Direct sowing without any field management practice. This was used as the control treatment.

6.2.2   Field lay out, cultivation practices and measurements

The experimental fields were ploughed and harrowed at the onset of the rainy season in both years. The stumps were removed manually before the setting and the marking out of the plots and after which the treatments were put in place. A Non-selective Systemic foliar-herbicide in the form of 480 g/L Glyphosate-Isopropy lamine, salt-force up^{TM} was applied on the prepared land at 3 litres ha^{-1} for the control of stubborn, annual and perennial grasses. The treatments were laid out in a randomized complete block design (RCBD) with four replicates (Figure 6.3). The cultivar planted was TGX 1448 2^E, an indeterminate variety obtained from IITA. The variety was selected for planting and the research because it has been found to be suitable for cultivation during the rainy season in the area (Smith, 2006).

The initial moisture content, temperature and electrical conductivity were measured with 5TE (Decagon Devices, Inc., WA USA), at a depth of 10 cm on the day of planting. The row and plant spacing were 0.6 by 0.3 m, which produced 55,556 plants ha^{-1}. The size of each plot was 6 m by 6 m (36 m^2). The plots were separated by an alleyway of 1 m and 2 m at the boundaries of the field in order to prevent insects and rodents from attacking the plants. Ditches of dimension 0.3 m deep and 0.3 m wide were constructed around the fields to divert runoff from the adjacent fields. Similarly, drains were constructed around each plot and linked together in order to ensure proper drainage and to prevent interference among the plots. Soybeans were planted on 24th May (DOY 144) in 2011 (first season) and 15 June, 2012 (DOY 167) (second season) barely 48 hrs after stabilization of the rainfall. The seeds were sown at a depth of 4 cm as recommended by Dugje et al. (2009) and Stedutos et al. (2012). Each plot was thinned to one plant per stand and mulches were applied to the selected plots.

Three rain gauges were positioned in the experimental fields in both seasons to measure daily rainfall. Soil moisture (sensors) meters (Decagon Devices, WA, USA) were installed at depths of 10 - 20; 20 - 30; 30 - 40, 40 - 50 and 50 - 60 cm in two replicates of each treatment and were used to monitor the moisture content and temperature of the soil during the growing cycle after they had been calibrated. Moisture contents were measured at intervals of 1 week with the assumption that the water requirements of the crops for a period of seven days would be catered for by water storage in the soil. Sampling of the moisture content was limited to 0.6 m depth, which has been described as the zone of greatest root density of Soybean in Obalum et al. (2011).

*Figure 6.2. Experimental fields showing the treatments and their lay out during the
rainfed experiments in 2011 and 2012*

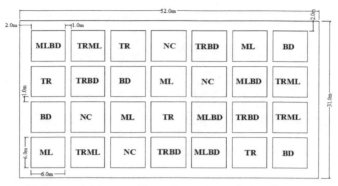

Figure 6.3. Lay out of the experimental treatments during the rainy seasons

At the beginning of vegetative growth, flowering and pod formation, insects and defoliating beetles observed on the field were controlled by using Lambda-Cyhalothrin 15 g\L+Dimethoate 300g\LE known as Magic Force[TM] (Jubaili Agro Chemicals) at 1.5 litre ha[-1]. Plants within an area of 1 m² at the centre of each plot were marked and sampled for plant heights, number of leaves and branches at intervals of seven days. The plant height was measured with a graduated metre rule. Soil water storage (mm) at vegetative (VE - V2) and reproductive stages namely: flowering FL (R1), pod initiation PI (R3), seed filling PF (R6) and maturity MT (R8), as reported by Burton (1997), were determined by using the equation expressed as (Ali, 2010; Liu et al., 2013):

$$SWS = \sum_{i=1}^{n} \theta_i \times z_i \tag{67}$$

where:
SWS = total soil water storage (mm)
θ_i = moisture content for soil layer i (m³ m⁻³)
z = soil depth or layer i (mm)
n = number of soil layers within the root zone

Soil water storage was determined for the upper 10 - 30 cm where about 90% of the active roots were located (Steduto et al., 2012). The values obtained at the stated stages of growth were added to obtain seasonal water storage.

After physiological maturity of the crop on 18 September, 2011 (DOY 261) at 117 DAP and 13 October, 2012 (DOY 287) at 120 DAP, plants within 13.4 m² at the central rows of each plot were harvested and threshed. The dry chaffs and stems were separated from the seeds and weighed. Their proportions (%) in each treatment were determined. Seed yields were measured by using a digital weighing balance (Kern model, PCB 3500-2; accuracy of 0.01 g; www.vosinstrumenten.nl). The seed yields per ha in each treatment were estimated. HI was determined from the ratio of the dry seed after harvest to (DAB) dry above ground biomass (Siahpoosh and Dehghanian, 2012).

6.2.3 Leaf area index and fraction of Intercepted Photosynthetically Active Radiation

At an average interval of 10 days from 18 DAP and 7 days from 28 DAP in the first and second seasons respectively, the LAIs, above and below PARs were measured by using an AccuPAR LP 80 (Decagon Devices, Inc., Pullman, USA) until physiological maturity. The instrument has a probe length of 86.2 cm and contains 80 sensors spaced at intervals of 1 cm along its length. Measurements were taken near noon and about 5 minutes were spent in each plot (Figure 6.5). Ten samples of the below and above PARs were taken from three replicates of each treatment by placing the probe (line sensor) perpendicularly to the rows above and below the plant canopy. The average value of LAIs was measured and the standard error of measurement computed for each of the treatments. A total of 11 and 13 consecutive measurements of LAIs were made in the first and second seasons respectively. The fraction of the IPAR (F_i) was determined in the two seasons by using the ratio of the PAR_{below} to the PAR_{above} as:

$$F_i = 1 - \left(PAR_{below} / PAR_{above} \right) \tag{68}$$

Seasonal λ was determined from the LAIs and their corresponding F_i with intercept set at zero because PAR_{above} is equal to the PAR_{below} when LAI = 0 (Robertson et al., 2001; Farahani et al., 2009; Narayanan et al., 2013). This is expressed by using the equation:

$$\lambda = -\ln(1 - F_i) / LAI \tag{69}$$

The slope of each graph was taken as the seasonal λ for each treatment (least square regression). The daily LAI for each treatment was determined by interpolation of the measured values. By using the daily LAI and the mean seasonal λ, the actual fIPAR of the plant for each day was computed based on the Ritchie type of equation (Ritchie, 1972; Ritchie et al., 1985) as applied by Farahani et al. (2009):

$$fIPAR = 1 - \exp(-\lambda \times LAI) \tag{70}$$

The above and below photo flux density (PFD) (μmol m^2 s^{-1}) was converted to energy flux (W m^{-2}) by using the conversion of 2.35×10^5 J mol^{-1} PARs (Campbell and Norman, 1998) and the cumulative intercepted PARs (TIPAR) from the inception of measurements until physiological maturity was determined for each treatment.

6.2.4 Dry above ground biomass

At intervals of two weeks from 21 days after planting (DAP) in the first season and one week from 28 DAP in the second seasons, the above ground biomass were taken from an area of 0.179 m^2 at the centre of the plots from four replicates of each treatment. During harvesting of the biomass, the growth stage was recorded. The harvested biomass was oven dried at a temperature of 70 °C for 48 hr until constant weight. The DAB per unit area was later estimated. A total of 11 and 13 consecutive weekly measurements of the above ground biomass was made in the first and second seasons. The cumulative DAB in each treatment was plotted against day after planting.

6.2.5 Radiation Use Efficiency

The proportion (f_e) of the PARs in SR (PAR/SR) was determined from continuous measurement of both SR and PARs for a period of 30 days when direct measurements of PAR were available at the weather station. Daily IPAR (MJ m^{-2} day^{-1}) from SR was determined by using the equation (Campbell and Norman, 1998; Narayanan et al.,

2013):

$$IPAR = f_e \times SR\left[(1 - \exp(-\lambda \times LAI)\right] \tag{71}$$

The daily IPARs were successively summed together to obtain the cumulative IPAR (TIPAR). TIPARs were also determined from instantaneous measurements of PARs from emergence until physiological maturity. Two approaches were used in determining the RUE of the crop. Firstly, fitting a linear function to the cumulative DAB and TIPARs (MJ m^{-2}) measured at intervals of one week (instantaneous measurement). Secondly, by fitting a linear curve to the cumulative DAB and TIPAR derived from the SR during the days the measurements of the biomass were made (Purcell et al., 2002; Tesfaye et al., 2006; Narayanan et al., 2013 and Ceotto et al., 2013). The slope of the linear function in each approach was taken as the RUE of the crop for each treatment. The RUEs for each treatment using the two approaches were compared by using linear regression and absolute differences between them were determined.

Equation 63 was used in simulating the accumulation of biomass throughout the growing cycle in each treatment. The measured and simulated biomasses were compared and evaluated by using the statistical indices r^2, NRMSE and d.

6.2.6 Crop transpiration

The crop coefficients at the mid and late seasons were adjusted by using the crop and weather data (Allen et al., 1998). The maximum daily evapotranspiration was determined by considering the rainfall frequency and depth, soil types and infiltration depths at the study site in the two seasons. The $K_{c\ mid}$ and $K_{c\ end}$ were adjusted by using the Equations 72 and 73:

$$K_{cmid} = K_{cmid(tab)} + \left[0.04(u_2 - 2) - 0.004(R_{H\min} - 45)\right] \times \left(\frac{h}{3}\right)^{0.3} \tag{72}$$

$$K_{cend} = K_{cend(tab)} + \left[0.04(u_2 - 2) - 0.004(R_{H\min} - 45)\right] \times \left(\frac{h}{3}\right)^{0.3} \tag{73}$$

where:
$K_{c\ mid\ (tab)}$ = crop coefficient for the middle stage (Table 12 of FAO-56 Bulletin)
$K_{c\ end\ (tab)}$ = crop coefficient for the final stage (Table 12 of FAO-56 Bulletin)
u_2 = average daily wind speed at 2.0 m height
h   = plant height (m) at mid- season and late stage of growth

Crop coefficients at initial stage ($K_{c\ ini}$), mid season ($K_{c\ mid}$) and end of season ($K_{c\ end}$) in 2011 were 0.94, 1.09 and 0.43, while in 2012 the comparative values were 0.80, 1.06 and 0.34. The potential (maximum) crop water use (ET_c) was determined by using the single crop coefficient approach as expressed by (Allen et al., 1998):

$$ET_c = ET_o \times K_c \tag{74}$$

Time series graphs of the fIPARs in each year were drawn for all the treatments. Transpiration at any period during the growing season was estimated from the product of the maximum possible crop water use (mm day^{-1}) and fIPAR expressed as (Ventura et al., 2001):

$$T_i = ET_c \times \left(\frac{fIPAR}{100} \right) \tag{75}$$

where:
T_i = transpiration (mm day^{-1})
ET_c = as defined previously
fIPAR = daily fraction of IPAR from emergence till the maturity

Seasonal transpiration (mm) was determined for each treatment by adding together the transpiration from emergence until maturity.

6.2.7 Actual crop evapotranspiration

The actual crop evapotranspiration was determined by using the soil water balance approach and expressed by the equation (Ali, 2010):

$$ET_c = P + I + CR - DP - RO \pm \Delta S \tag{76}$$

where:
P   = daily precipitation (mm)
I   = irrigation depth (mm)
CR   = capillary rise (mm)
DP   = deep percolation (mm)
RO   = runoff (mm)
ΔS   = change in water content (mm)

In the two rainy seasons, rainfall was adequate for the cultivation of the crop, so no irrigation water was applied. The contribution of the groundwater was ignored because the groundwater table was deeper than 20 m at the experimental fields and therefore capillary rise was ignored. Runoff was measured from ML and NC by installing rigid metallic boxes (50 × 50 × 80 cm) around plants within an area of 0.716 m^2 in the two replicates. Runoff within the area was channelled towards graduated plastics and measured immediately after each rainfall event (Araya, 2010). Surface runoff in other treatments was considered negligible because the heights of the Soil bunds and ridges were up to 30 cm (Ali, 2010) and no case of bund or ridge overflow was observed. Change in water content or storage (Δs) was determined from the difference between moisture storage on all the sampling dates and their respective preceding dates. Percolation below the root zone ($\leq$ 60 cm) was determined from the soil moisture contents measured periodically while for deeper depths beyond 100 cm, it was assumed to be negligible (Ali, 2010). The equation is therefore reduced to:

$$ET_c = P - RO - D \pm \Delta S \tag{77}$$

Seasonal soil evaporation for each treatment was determined by subtracting the seasonal transpiration (mm) from the actual crop evapotranspiration (mm). Water productivity (kg ha^{-1} mm^{-1}) was determined by using Equation 60 while economic output was determined from the product of water productivity and price of marketable seed yields. Transpiration efficiency (TE) was determined from the ratio of the DAB at harvest (kg ha^{-1}) to the seasonal transpiration (mm).

6.2.8 Field observations of plant phenologic development in the rainy season of 2011

The season was divided into four growth periods namely:
- *initial stage (VE-V2)*. The seeds were planted on 24 May, 2011 and 90% emergence was recorded on 3 June, 2011 (10 DAP). A 10% canopy cover, which marked the end of the initial stage was recorded on 18 June, 2011 (25 DAP). The initial stage lasted for 25 days;
- *development/vegetative stage*. The development stage (V1-V6) was characterized by leaf development and expansion, increment in plant height and this started on 19 June, 2011 (26 DAP) and ended on 27 July, 2011 (64 DAP), it lasted for 38 days;
- *flowering began (R1-R2) on 3 July, 2011 (40 DAP) and ended on 23 July, 2011 (60 DAP), it lasted for 20 days.* Flowering was observed in almost all the treatments at the same time. However, not all plants in the plots (treatment) flowered at the same rate. There was variability in the degrees of flowering among plant stands in the same plot, which is similar to the observations of Egli (1994). The variety (cultivar) is indeterminate because the production of more leaves continued after flowering;
- *mid-season*. Characterized by pod initiation (R3) to full blown pod (R4), seed filling (R5) and full seed (R6). Mid-season (R3-R6) began on 24 July, 2011 (60 DAP) and ended on 30 August, 2011 (98 DAP). Mid season lasted for 38 days;
- *late season*. Characterized by the end of seed formation in pods, gradual shedding of leaves (senescence), and commencement of seed maturity and drying of the pods (R7-R8). Beginning of maturity (R7) started on 31 August, 2011 (99 DAP) but ended on 18 September, 2011 (117 DAP) after full maturity (R8). It lasted for 18 days. The seeds were harvested two day after physiological maturity.

6.2.9 Field observations of plant phenologic development in the rainy season of 2012

The growing season was divided into four growth periods namely: initial, development, mid-season and late stage. Overlapping of the growth stages and extension of days were observed among the plots (Figure 6.4):
- *initial stage*. The initial stage (VE-V2) ran from planting to 10% ground cover by canopy (Allen et al., 1998). The seed was planted on 15 June, 2012 and 90% emergence was observed and recorded on 25 June, 2012 (10 DAP). The 10% cc_o was recorded on 9 Jully, 2012 (24 DAP) which marked the end of the initial stage. Therefore the initial stage lasted for a period of 24 days after planting;
- *development/vegetative stage*. The development stage (V1-V6) was characterized by foliar development, expansion and increment in plant height. Development stage began on 10 July, 2012 (25 DAP) and ended on 14 August, 2012 (60 DAP). It lasted for 35 days. More leaves and increment in plant height were recorded until 14 September, 2012 (91DAP);

Figure 6.4. Experimental field during rainfed conditions showing: (a) ploughed land in preparation for the cultivation; (b) measurement of plant biometrics; (c and d) sampling of LAIs by using a Ceptometer; (e) Soybean during the seed filling stage; (f) Senescence at late stage; (g and h) late stage when about 70% of the pods were ripe and matured

- *flowering began (R1-R2)* on 29 July, 2012 (44 DAP) and ended that is full blown flowering (R2) on 14 August, 2012 (60 DAP), it lasted for 16 days. Flowering (R1) was observed in almost all the treatments at the same time. However, not all plants in each plot (treatment) flowered at the same rate. There was variability in the degrees and duration of flowering among plant stands in the same plot and for different treatments;
- *mid-season.* The mid season is characterized by initiation of pods (R3) to full blown pods (R4) and commencement of seed filling (R5). Pod initiation started on 12 August, 2012 (58 DAP) and ended on 18 August, 2012 (64 DAP); it lasted for 7 days. Seed filling (R5) began on 19 August, 2012 (65 DAP) and ended on 24 September, 2012 (101DAP) with the pods completely filled up with seed (R6), it lasted for 37 days. Both the pod initiation and seed filling lasted for 44 days. The mid-season stage is the longest stage for perennials and for many annuals (from Allen et al., 1998);

- *late season*. The late season was characterized by senescence that is, yellowing, gradual shedding of leaves and the beginning of maturity (R7 growth stage). Beginning of maturity (R7) started on 25 September, 2012 (102 DAP) but ended on 13 October, 2012 (120 DAP) after full maturity (R8). It lasted for 18 days. The crop was harvested a day after physiological maturity (Figure 6.4).

6.3   Experimental treatments, field lay out, cultivation and measurements in the dry seasons

In the dry seasons of 2013 (February - May) and 2013/2014 (November, 2013 - February, 2014), Soybeans were also cultivated at TRFOAU. The field used for the dry season experiment was located at about 1 km away from the field used during the rainy seasons due to the nearness to the source of water (Figure 6.1). The treatments, field lay out, cultivation and measurements are stated in the sections that follow.

6.3.1   Experimental treatments

The experimental factor is application of water (irrigation). The treatments were:
- T_{1111} Irrigation was done weekly without skipping at any of the growth stages: flowering [beginning bloom (R1) and full bloom (R2)], pod initiation [beginning pod (R3) and full pod (R4)], seed filling [beginning seed (R5) and full seed (R6)] and maturity [beginning maturity (R7) and full maturity (R8)] (reference Treatment);
- T_{0111} Irrigation was skipped for seven days every other week during flowering only while weekly irrigation was observed at pod initiation, seed filling and maturity;
- T_{1011} Irrigation was skipped for seven days every other week at pod initiation only while weekly irrigation was observed at flowering, seed filling and maturity;
- T_{1101} Irrigation was skipped for seven days every other week during the seed filling stage only while weekly irrigation was observed during flowering, pod initiation and maturity;
- T_{1110} Irrigation was skipped for seven days every other week during maturity only while weekly irrigation was observed during flowering, pod initiation and seed filling.

6.3.2   Field lay out, cultivation and measurements in the dry seasons

The experimental field was harrowed at the commencement of the fieldwork in 2013. The stumps were removed manually before the setting out of the plots and after which the treatments were put in place. A Non-selective Systemic foliar herbicide in the form of 480 g/L Glyphosate-Isopropylamine, salt-force up^{TM} was applied on the prepared land at 3 litres/ha for the control of stubborn grasses. The treatments were laid out in a randomized complete block design with three replicates (Figure 6.4). Pre-wetting of the field was done to a depth of 20 mm in order to initiate germination of the seed. Soybeans were planted on 2 February, 2013 (DOY 33) (first season) and 8 November, 2013 (DOY 312) (second season). The difference in the time of planting was due to the variability and fluctuations in the weather conditions of the area. The seeds were sown at a depth of 4 cm. The cultivar planted was TGx 1448 2^E, the indeterminate variety obtained from IITA. It was selected in order to compare the yield components with those obtained during the rainy seasons (sub-section 6.2.2). Insects and defoliating beetles were observed and controlled by using Lambda-Cyhalothrin 15g\L+Dimethoate 300g\L E known as Magic Force™ (Jubaili Agro Chemicals) at 1.5 litre ha[-1] at intervals of two weeks. Three seedlings were thinned to one plant per stand at 25 DAP after full establishment in each season. A pressure compensating inline-drip line (Drip works, USA) whose capacity was 2.2 l h[-1] with operating pressure of 1 bar was used in applying water. The length of each lateral was 5 m and contained 17 point in-line emitters, which were pre-spaced at intervals of 0.3 m. The plant spacing was 0.6 by 0.3 m, which produced 55,556 plants ha[-1]. Each plot contained 68 plants (9 m[2]) arranged in four rows that is, 17 plants per row on flat land (Figure 6.5). An alleyway of 1 m was used to separate the plots and to allow for easy movement. The entire area of the experimental field was 19 m by 15 m (285 m[2]). The total area occupied by plants was 135 m[2].

Moisture contents in the soil layers were measured by using the gravimetric method from two replicates while available soil moisture (mm) prior to irrigation was determined from the product of soil moisture content (g g[-1]), bulk density (g cm[-3]) and the average depth of the soil in the root zone. Plants within an area of 1 m[2] at the centre of each plot (treatment) were marked and sampled for plant height, number of the leaves and branches at seven days intervals throughout the growing season in the three replicates. The plant height was measured with a graduated metre rule from the soil surface to the point of emergence of new leaves. After maturity on 25 May, 2013 (DOY 145) in 112 DAP and 25 February, 2014 (DOY 56), 110 DAP (sub-sections 6.3.5 and 6.3.6), an area of 5.37 m[2] in the central rows was harvested from each of the plots and the seed yields per ha were estimated. HI was determined (sub-section 6.2.2).

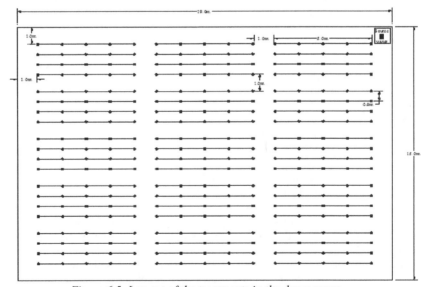

Figure 6.5. Lay out of the treatments in the dry seasons

Design of the drip irrigation system

The crop water use was estimated by using the single crop coefficient approach (Allen et al., 1998). The length of each stage was taken from the records taken during experiments conducted in the rainy seasons (sub-sections 6.2.6 and 6.2.7). The estimated peak evapotranspiration during the initial, and development stages were 1.13 and 6.53 mm day^{-1} respectively, while at mid and late stages of growth they were 6.69 and 3.83 mm day^{-1} respectively.

Selection of emitters

The emitter selected for this experiment was the point source in-line emitter (Dripworks, Inc., USA). It was equipped with pressure compensating mechanisms, which ensure even distribution of pressure along the laterals even in hilly and undulating areas (Michael, 2008). At commencement of the experiments, the coefficient of variation of the discharges from the emitters was determined by using the expression:

$$CV = \frac{\sqrt{\dfrac{\sum\limits_{x=1}^{n}(Q_x - Q_{av})^2}{n-1}}}{Q_{av}} \qquad (78)$$

where:
Q$_x$ = discharge of each emitter under analysis (l h^{-1})
Q$_{av}$ = average emitter discharge (l h^{-1})
N   = number of observations

The CV was 0.03 and is excellent (Michael, 2008) for the point source emitter. The statistical uniformity indicator U$_s$ was 95%. The emission uniformity of the drip system of 90.7% was estimated from the equation of Keller and Bliesner (1990) expressed as:

$$EU = 100 \times \left(1 - 1.27 \times \frac{CV}{\sqrt{n}}\right) \times \frac{Q_{min}}{Q_{av}} \qquad (79)$$

where:
EU   = emission uniformity
n   = number of emitters per plant (1)
Q$_{min}$ = minimum emitter discharge along the laterals (l h^{-1}) computed from the minimum pressure along the laterals that is 1.80 (l h^{-1})
Q$_{av}$ = as previously defined and is 1.91 (l h^{-1})

Volume of water required per plant per day at the initial stage was determined from the ratio of the product of peak evapotranspiration (1.13 mm day^{-1}) and wetted area of 0.054 m^2 (30% of the area of 0.179 m^2) occupied by each plant to the emission uniformity of 90.7% was 0.06 litres. The initial stage was expected to last for 25 days under the conditions in Ile-Ife and the estimated total volume of water to be supplied to each plant was 1.5 litres (25 days). The estimated field water requirement at initial stage was 1,530 litres (17×4×15×0.06×25). The daily water requirements at the mid (40 days) and late (18 days) seasons were 6.69 and 3.83 mm day^{-1}. By using the same procedure, the estimated daily water needs per plant during these stages were 0.36 and 0.21 litres respectively. Similarly, the amounts of water budgeted for the entire experimental fields during these periods were 14,700 and 3,860 litres respectively. The readily available moisture at the root zone (average of 10 cm) at the initial stage was determined from Equation 80:

$$RAM = WHC \times RZD \times p \qquad (80)$$

where:
RAM = readily available water (mm)
WHC = water holding capacity (110 mm m^{-1} obtained from field survey)
RZD = root zone depth (10 cm at initial stage)
p = maximum allowable soil moisture depletion (fraction = 0.5)

At initial stage the RAM was 5.5 (6 mm). Irrigation frequency was determined from the ratio of the RAM to peak water use of 1.13 mm day^{-1}. This gave an average of 4.98 days (5 days). From the calibrated flow rate of the emitter and wetted area of the plants, adding 5.5 mm of water per irrigation would take 0.2 hours (12 minutes). Daily effective rainfall was determined by using the equation (Koegelenberg, 2003):

$$R_e = R - 1.5 \times E_p \qquad (81)$$

where:
R_e = effective rainfall (mm)
R = daily rainfall (mm)
E_p = pan evaporation (mm)

Measurement of soil moisture

During the irrigation seasons, wet soil samples were collected by using a 53 mm internal diameter steel core sampler from two replicates of each treatment at intervals of 10 cm from 0 to 60 cm and the moisture content was determined with the gravimetric method. The samples were weighed immediately on the field, kept in sealed polythene bags before transported to the laboratory. The samples were oven-dried at 105 °C for about 48 hrs until constant weight. The volumetric water content in linear depth (mm) was determined by multiplying soil moisture measurements (%) by bulk density of each layer and depth of the root zone. There was rainfall in a few days during the fieldwork and this was built into the irrigation schedule by adding the effective rainfall to the plant available water and computing the number of days it would take plant to use it.

Soil around the roots was carefully excavated, the roots were washed and measured on millimetre paper in order to determine the root depth (Ali, 2010). Average root depth at intervals of 1 week was used to schedule the irrigation at initial, development, mid and late seasons. At the initial stage, irrigation was done to a root depth of 10 cm, while at development and mid seasons, irrigation was done to a depth of 30 cm. The percentage depletion of the available soil moisture in the effective root zone of the plant was determined with the following equation (Ali, 2010):

$$Depletion\,(\%) = \frac{1}{n} \sum_{i=1}^{n} \frac{M_{fci} - M_{bi}}{M_{fci} - M_{pw}} \qquad (82)$$

where:

M_{fci} = field capacity in the ith layer of the soil (m^3 m^{-3})
M_{bi} = moisture content before irrigation in the ith soil layer (m^3 m^{-3})
M_{pw} = moisture content at permanent wilting point (m^3 m^{-3})

For each treatment, irrigation water was equally applied (same amount and frequency) until the commencement of flowering when skipping of irrigation began in the selected treatments (sub-section 6.3.1). The water holding capacity of the soil was 110 mm m^{-1}. The net irrigation requirement of the crop that is, the amount of water applied during each stage of growth was determined by using Equation 83 and expressed as:

$$d = \sum_{i=1}^{n} \frac{(M_{fci} - M_{bi})}{100} \times A_i \times D_i \qquad (83)$$

where:

d = net amount of irrigation applied (mm)
M_{fci} = field capacity in the ith soil layer (m^3 m^{-3}). It was measured two days after irrigation
M_{bi} = moisture content before irrigation in the ith soil layer (m^3 m^{-3})
A_i = bulk density of the soil in the ith layer (g cm^{-3})
D_i = depth of the ith soil layer within the root zone (mm)
n   = number of soil layers in the root zone

Irrigation frequency at each stage was determined from the ratio of the net water requirement (mm) to the peak water use (mm day^{-1}). The area irrigated by each dripper was determined from the ratio of the product of plant spacing and percentage of the cropped area irrigated to the number of drippers at each emission. Only 30% of the cropped area was irrigated.

Leaf area index

At average intervals of seven days from 14 DAP in both irrigation seasons, the green LAI, above and below PARs were measured by using an AccuPAR LP 80 (Decagon Devices, Inc., WA, USA) until maturity of the seeds. The procedures used in taking the measurements during the dry season were the same as stated in sub-section 6.2.3. Total of 14 consecutive measurements of LAIs were made in both seasons. The fIPAR for each treatment was determined in the two seasons by using Equation 70.

6.3.3 Measurement of soil water evaporation

Evaporation from the free water surface was measured with a class (A) pan installed on the experimental field on a daily basis throughout the crop cycle. A time series graph of LAI versus day after planting was developed from which the LAI of the crop at any period in each treatment was determined. Assuming that the net radiation inside a canopy decreases according to an exponential function and that the soil heat flux is neglected, daily actual evaporation of water from the soil in each treatment was determined by using the method of Cooper et al. (1983); Belmans et al. (1983) and Lu et al. (2011) which is expressed as:

$$E_a = EXP(-\lambda \times LAI) \times E_p \tag{84}$$

where:
E_a = actual evaporation from the soil in a cropped plot (mm)
λ = seasonal leaf extinction coefficient = 0.46 (-)
LAI   = leaf area index ($m^2 m^{-2}$)
E_p = pan evaporation by using a class A pan (mm)

Seasonal soil water evaporation was determined by summing daily evaporation from emergence until physiological maturity.

6.3.4 Actual crop evapotranspiration in the dry season

The actual crop evapotranspiration was determined by using Equation 76. Daily rainfall was measured with three rain gauges that were installed on the field. Moisture content was measured before irrigation in order to refill the soil at the root zone to field capacity. Runoff was measured by placing metallic boxes (50×50×80 cm) around plants within an area of 0.716 m^2 in two replicates. The runoff within the area was directed towards a graduated plastic and measured after each rainfall event (Araya, 2010). The contribution of the groundwater was ignored because the groundwater table was deeper than 10 m. The change in the moisture ($\pm\Delta s$) at the root zone was measured at the beginning and end of each stage of growth of the crop. The water was applied to replace the soil moisture in the root zone, therefore the drainage below the root zone was considered negligible and ignored under drip irrigation because water was applied to replace the soil moisture in the root zone (Lovelli et al., 2007). Equation 76 was therefore reduced to:

$$ET_c = I + P - R_o \pm \Delta S \tag{85}$$

Crop water uses at different stages were aggregated to obtain seasonal water. Seasonal transpiration was determined from the difference between seasonal actual crop water use and evaporation of water from the soil (Lu et al., 2011). Crop water productivity and irrigation water productivity were determined by using Equations 60 and 61 respectively. Economic water productivity was determined from the product of the price and crop water productivity (Equation 62).

6.3.5 Dry above ground biomass under water deficit conditions

At intervals of seven days from 14 days after planting (DAP) in both irrigation seasons, the above ground biomass were obtained from an area of 0.179 m^2 at the centre of the plots from two replicates of each treatment. The harvested biomass was oven dried at 70 °C for 48 hr until constant weight. The DAB per unit area in each treatment was estimated. Total of 14 consecutive weekly measurements of the DAB were made in both irrigation seasons.

6.3.6 Field observations of plant phenologic development in the dry season of 2013

The duration of the crop growth was sectioned into four stages similar to those stated in Allen et al. (1998) and these are establishment (initial), development (vegetative), mid season (flowering-pod initiation) and late season (seed filling and maturity). Each of these stages is identified based on agronomic features of the crop, which are monitored from planting until harvest. Overlapping was observed in few of the growth stages:

- *establishment (initial) stage.* The seeds were planted on 2 February, 2013 (DOY 33). A 90% germination of the seed was observed on 12 February 2013 (10 DAP). A 10% canopy cover was observed on 27 February (25 DAP) which marked the end of the initial stage/establishment of the crop in 2013. The initial stage is represented by (VE-V2) where VE is the emergence, V1 is the first trifoliate, V2 is appearance of the second trifoliate and the stage at which active nitrogen fixation has began;

- *development stage.* The development stage began on 28 February, 2013 (26 DAP) and the canopy cover was measured on this day in the different treatments. The development stage is characterized by leaf expansion and formation of more canopies. The cultivar under study is indeterminate and as a result, foliar expansion extended until the beginning of the mid season;

- *flowering.* R1 growth stage, that is beginning bloom, began on 17 March, 2013 (43 DAP) and continues through R2 (full bloom growth stage) but ended on 29 March, 2013 (55 DAP). Flowering lasted for 12 days. Irrigation was skipped for only seven days during the flowering period in T_{0111};

- *mid-season.* Starting on 30 March, 2013 (56 DAP) with pod initiation (R3 growth stage). Pod initiation (R3 growth stage) commenced on 30 March, 2013 (56 DAP) but ended on 6 April (63 DAP), it lasted for 7 days. During this period, irrigation was skipped for the entire seven days in T_{1011}. Full pods (R4 growth stage) were observed on 7 April (64 DAP). Seed filling (R5 growth stage) commenced on 8 April (65 DAP) and skipping of irrigation commenced at this stage of growth of the crop (T_{1101}). The complete filling of pods (R6) with seeds ended on 14 May, 2013 (101DAP). Seed filling lasted for 36 days and the entire mid season lasted for 43 days;

- *late season.* Characterized by senescence and commencement of maturity of the crop began on 15 May, 2013 (102 DAP) and ended on 25 May (112 DAP). The crop was harvested a day after maturity.

Seed filling of Soybeans is mostly taken as the period from commencement of seed filling (R5) to the time of full seed (R7). This approach was used in this study (Egli et al., 1984; Agudelo et al., 1986).

6.3.7 Field observations of the plant phenologic development in the dry season of 2013/2014

The following observations and records of phenologic development of the crop were made in the second season:

- *establishment (initial) stage.* The seeds were planted on 8 November, 2013 (DOY 312). A 90% germination of the seed was observed on 16 November (8 DAP). A 10% CC was observed on 24 November (16 DAP);

- *development stage.* The development stage began on 23 November, 2013 (15 DAP). The canopy cover was measured on this day in the different treatments.

The development stage is characterized by leaf expansion and formation of more canopies;

- *flowering.* R1 growth stage began on 16 December, 2013 (38 DAP), continued through R2 (full bloom growth stage) and ended on 27 December (49 DAP). Flowering lasted for 12 days. Irrigation was skipped for only seven days during the flowering period in T_{0111};

- *mid-season.* Started on 28 December, 2013 (50 DAP) with pod initiation (R3 growth stage). Pod initiation commenced on 28 December (50 DAP) and ended on 4 January, 2014 (57DAP), it lasted for 8 days. Full pods (R4 growth stage) were observed on 5 January (58 DAP). During this period, irrigation was skipped for the seven days (T_{1011}). Seed filling (R5 growth stage) started on 6 January (59 DAP) and skipping of irrigation (T_{1101}) commenced. Seed filling ended on 15 February (99 DAP). It lasted for 31 days. Seed filling lasted for 40 days and the entire mid-season lasted for 50 days;

- *late season.* Commenced on 18 February, 2014 (102 DAP). Skipping of irrigation began at commencement of maturity (R7) in (T_{1110}). The crop reached full maturity (R8) on 25 February (109 DAP). The late season lasted for 7 days and the crop was harvested 12 hours after physiological maturity. The duration of the growing season from planting until physiological maturity was 110 days.

Table 6.1 contains the summary of the duration of the five growth stages during the rainy and dry seasons. In 2011 and 2012, seed filling constituted about 33 and 35% of the entire growing seasons. The duration of each stage in this study falls within the 21 - 46% recorded for Soybeans (Egli, 1994; Steduto et al., 2012).

Table 6.1. Summary of the duration (days) of the key phonologic stages of the crop in each year

Year	Planting to 90% emergence	Flowering	Pod initiation	Seed filling	Maturity
Rainfed conditions					
2011	10	20	6	32	18
2012	10	16	7	37	18
Mean	10	18	6	34	18
Irrigated conditions					
2013	10	12	7	36	10
2013/2014	9	12	8	46	07
Mean	10	12	8	41	09

6.4   Economic analysis

The economic analysis was done for the two seasons in order to know the profitability of using drip irrigation in the cultivation of the crop. In the study area, water is considered to be the limiting resource because of fluctuation in rainfall during the rainy seasons. Increasing urbanization and land degradation may reduce agricultural land and thereby creating scenarios of land limiting situations. Analyses of both land and water limiting conditions were done by following English et al. (1990).

6.4.1   Land limiting conditions

Under land limiting conditions, the optimum practice is the one that maximises the net return per unit of land. In this study, the cost of production under rainfed conditions consisted of operation cost, which concern seed, labour, insecticides, weeding, repair of bunds and ridges and processing after harvesting (Appendix L). The cost of mulch and hiring of land were not factored into the production cost because mulch is available free of charge and most farmers own the land by inheritance. The gross revenue was determined by multiplying the average seed yield (land productivity) by the annual average price of the crop in the two seasons at the international market (FAO, 2014). The net financial return/loss was determined by subtracting the total cost of production from the total revenue. The seed yield, cost and revenue per hectare was plotted as a function of seasonal crop water use.

Under full and deficit irrigation conditions, the fixed and variable items considered in the economic analysis are shown in Appendix P. Two scenarios were considered in the determination of the cost of production. First, a situation whereby the water applied to the crop is costed and included in the total cost of production. Second, a situation whereby the water used for irrigation either is provided without cost by an agency of the government or is pumped from a stream without cost, except for the fuel for the pump as it was done in this study. The fixed cost of production was spread over a period of ten years in the two scenarios with the assumption that it can be refunded within the period. The gross revenue and financial return/loss were determined by using the approach stated earlier. The seasonal crop water use was plotted against average seed yield and total revenue.

6.4.2 Water limiting conditions

Under the water limiting situation, the cultivation practice that produces the optimum economic water productivity is considered to be the most promising method. The economic water productivity of the conservation practices was compared with the conventional practice. Economic analysis under water limiting conditions was also done for the dry season experiment. Water productivity and irrigation water productivity were compared. Number of irrigations was plotted against production cost and economic water productivity.

6.5   Statistical analysis

6.5.1   Rainfed conditions

The statistical software SAS was used for the analysis. The plant heights, number of leaves, LAIs of the crop, the fIPAR, accumulated DAB, yields and HIs from each treatment were analysed by using Anova (Duncan Multiple Range Test) at a significant level of $\alpha = 0.05$. Values for each replicate were averaged and the standard error of each treatment was calculated. Regression analyses were performed by using Sigma Plot 12.5. The graphs were draw by using Sigma Plot 12.5 and Origin 7.0 softwares.

6.5.2   Irrigated conditions

The analyses of variance of LAIs, DAB, yields and HIs were done by using SAS software at $\alpha = 0.05$. Averages of each parameter were determined and the standard error was computed. Student's t test at $\alpha = 0.05$ was used in comparing the means of the seasonal yields, WPs and IWPs for the two seasons. Regression analyses were performed with Sigma Plot 12.5. The graphs were drawn with Sigma Plot 12.5 and Origin 7.0 softwares.

7 Water conservation, soil water balance and productivity of Soybeans under rainfed conditions

Potential yields of many crops, which are obtained or generated by simulations, are available in literature. However, in most areas in the developing countries, farmer's yields are significantly lower than the potential yields. This is due to poor or inefficient crop and water management options at farm scales under rainfed systems. A vast potential of rainfed systems needs to be harnessed through careful management of the natural resource to increase productivity and achieve food security in the developing countries. In order to double the world food production, there is an urgent need to adopt soil and water conservation techniques that are cost effective and easy to implement at farm scales. It has been identified that a core problem, which water saving agriculture has to solve is how to raise the water utilization rate and water productivity under rainfed and irrigated conditions (Waraich et al., 2011). Management of soil and water at farm scales in rainfed areas in the developing countries will play a critical role in achieving the full potential (Wani et al., 2009). Water needs to be harvested during periods of high rainfall, stored in the soil and re-used during dry spells. Many methods of water conservation techniques are reported in literature. These include deep tillage, mulching, broad based beds and furrows, ridges and furrows and compartmental bunding (Muthamilselvan et al., 2006). Different types of materials such as straw, rice straw or husk, plastic film, grass, wood, sand and oil layer are used as mulch (Khurshid et al., 2006; Seyfi and Rashidi, 2007). Water productivity and weed control on farmland can be increased with mulch (Unger and Jones, 1981). Mulch (ML) has been widely used to increase the water intake and storage of soils (Schneider and Mathers, 1970), to improve moisture distribution in soil profiles and reduce evaporation (Bennett et al., 1966). Mulch improves root growth in the upper 15 cm of the soil. Reduction in water evaporation in the upper 15 cm was reported to be very beneficial to crop growth as an addition to water intake in the soil (Chaudhary and Prihar, 1974; Khan, 1996). The use of mulch provides a more conducive environment for seed germination, moderates soil temperature, increases soil porosity and water infiltration especially during rainfall of high intensity. It reduces runoff and thereby controls erosion and suppresses the growth of weed. Muthamilselvan et al. (2006) reported that strip-tillage cabbage from a heavy mulch (ML) treatment was 56% higher than that from a stubble mulch treatment under cold conditions. They also reported that the use of mulches (ML) with crop residues increased moisture conservation and resulted in greater grain yields compared with other tillage practices. Mulches increase the yields of Maize by 17% while the use of straw mulch increased it by 52.1%. Application of straw mulches and soil tillage improve crop growth and yields of Maize (Bana et al. 2013). In a study focused on application of plastic-covered ridges and bare furrows, Ren et al. (2008) showed that with different levels of rainfall (230, 340, and 440 mm); ridge and furrow rainfall harvesting led to increase in the Maize yields of 82.8, 43.4 and 11.2% compared with conventional flat cultivation, while crop water use efficiencies were increased by 77.4, 43.1, and 9.5%, respectively. Improvement in water use efficiency in rainfed farming can be achieved by simple water harvesting systems and improved crop rotation among others.

Past management practices focused on surface and groundwater. Future management should focus relatively more on utilization of rainfall through improvement in rainfed agriculture. Combining agronomic and water conservation measures can reduce considerably the amount of water required to produce a crop (Jury and Vaux, 2007). Improved land management is very essential in increasing water productivity under rainfed systems. A large proportion of rainwater is lost through evaporation or runoff. Practices that enhance infiltration and storage capacity of soils will increase water available for productive transpiration and crop yield (IWMI, 2006).

In Ogun-Osun River Basin for instance, many land areas for agricultural production are located on hillsides with steep slopes. This can be found in places like Ilesha, Erin-Oke and other places in the Ijesha zone of Osun State. Erosion in these areas will reduce soil nutrients. When there is shortage of rainfall during rainy seasons, yields of crops fall and farmers incomes are reduced. Supplemental irrigation and conservation techniques need to be adopted in order to sustain crop production in these areas. Supplemental irrigation can be very expensive for the peasant farmers in these areas because the returns on their investments may not be justified. Management of the existing land and water resources on a productive and sustainable basis in these areas of Ogun-Osun River Basin is essential. This can be achieved by the use of water conservation practices in the cultivation of crops that can replenish soil fertility such as Soybean. Soil water conservation techniques, which can be used, include water harvesting in plots, tillage practices, mulching, soil bunds, terracing and combinations of these techniques. The use of these technologies if well managed and sustained can significantly reduce the water risk under the current climate change and lead to substantial increase in crop yields under rainfed farming in the basin. The results of the use of water conservation practices in the cultivation of Soybean under rainfed conditions described in section 6.2 are stated in the sections that follow.

7.1   Environmental conditions in the rainy seasons

Table 7.1 shows the meteorological data measured at the automatic weather station. The first season was characterized by a slightly higher temperature condition - average seasonal maximum and minimum temperatures of 32.8 and 20.2 °C, respectively - than the second - average seasonal maximum and minimum temperatures of 30.9 and 20.2°C, respectively. The second season was more humid - average seasonal maximum and minimum relative of 95.8 and 53.0%, respectively - than the first - average maximum and minimum relative humidity of 95.0 and 46.9%, respectively. The rainfall depths during the experiments in 2011 and 2012 were 539 and 761 mm respectively (Table 7.1). The experimental field in 2012 is located at about 500 metre away from the field used in 2011. In 2011, rainfall depth during reproductive stage in the month of August was 20% of the seasonal rainfall unlike in 2012 when the comparative rainfall depth was 34% at the same stage (Table 7.1). The second season experienced higher rainfall than the first season. The recession in rainfall commonly called the August break was higher in the second season than in the first (Table 7.1). In the first season, only 10% of the rainfall occurred during the initial and mid stages, while 16% occurred during the late stage. In the second season, about 38% of the rainfall occurred during the development stage (foliar expansion) of the plant, while about 11% occurred during the late season characterized by seed maturity (R7 - R8) and drying. Monthly coefficients of variation of rainfall were 63 and 62% respectively in 2011 and 2012. The rainfall patterns during the plant cycles in the years favoured the cultivation of the plant because lesser rainfall is required at the late stage of the plant unlike during the flowering and seed filling whereby deficit in rainfall can trigger water stress and causes substantial reduction in yields and its components.

Water conservation, soil water balance and productivity of Soybeans under rainfed conditions

Table 7.1. Meteorological data measured at the experimental fields during the rainy seasons (standard deviations in parenthesis)

Year/Month		Temperature (°C)			Relative Humidity (%)			Global Solar Radiation (MJ m⁻² day⁻¹)		Rainfall (mm)
		Max	Min	Mean	Max	Min	Mean	Max	Mean	Mean
2011	May	32.3	20.7	26.7 (2.8)	95.2	43.5	79.8 (12.3)	1077	172 (253)	20
	Jun	31.0	20.1	25.4 (2.5)	94.5	28.2	82.7 (10.6)	1089	156 (241)	173
	Jul	29.4	20.1	23.8 (1.9)	94.9	55.4	84.5 (07.8)	957	125 (182)	174
	Aug	39.5	20.1	23.7 (2.9)	95.2	55.3	86.3 (08.0)	1058	117 (201)	108
	Sep	31.7	19.9	24.5 (2.2)	95.3	52.2	62.2 (09.4)	1228	144 (244)	64
2012	Jun	32.3	20.1	24.9 (2.4)	95.1	51.0	83.6 (10.3)	981	162 (222)	126
	Jul	32.3	20.0	23.8 (1.8)	98.0	51.0	86.9 (07.3)	973	120 (175)	242
	Aug	29.3	20.0	23.2 (1.6)	95.1	51.0	87.4 (06.9)	965	101 (145)	43
	Sep	29.6	20.8	24.0 (1.8)	95.4	59.9	87.1 (08.0)	1003	117 (177)	259
	Oct	31.0	20.2	24.7 (2.6)	95.6	52.3	54.4 (10.7)	1003	160 (236)	91

The mean incident SR in the first season was $1081Wm^{-2}$ while in the second season it was 985 Wm^{-2}. Air temperature of the study area in the dry season (December/January-April) is generally higher than the temperature during the rainy season (May/June-November). Monthly meteorological data from 2010 to 2014 are shown in Appendix C.

7.2   Physical and chemical properties of the soil

The physical and chemical properties of the soil at the experimental fields in 2011 and 2012 after ploughing are shown in Table 7.2. Although the experimental fields are in the same area, variability occurs in terms of soil fertility and physical structure, and these may influence root penetration and affect yields under different treatments. Soil samples were collected at depths of 00 - 50 cm in each year from four plots (replicates) of each treatment. The field used in 2011 during the rainfed experiment is categorized as alfisol (Adekalu and Okunade, 2008).The alfisol is characterized sandy and clay soils and contains large quantities of ferruginous concretions and fragments of ironstone or overlaying massive ironstone pan (Vinvine et al., 1954). The 00 - 10 cm was characterized by sandy clay loam while 10 to 60 cm is sandy loam. The bulk density ranged from 1.45 to 1.60 g cm^{-3} from 00 to 30 cm while from 40 to 60 cm it ranged from 1.58 to 1.62 g cm^{-3}. Organic matter (OM) was higher at the upper 20 cm ranging from 1.95 to 2.04% while at lower depth of 60 cm it was 0.64%. Higher organic matter at the upper soil layer is attributable to accumulation of plant material over years due to the abandonment of the site. Soil pH ranged from 6.3 to 6.5 between 00 to 60 cm of the soil layer, which is adequate for the cultivation of Soybeans. Potassium ranged from 0.15 to 0.21 c mol kg^{-1} within the upper 30 cm while in the lower layer that is, 30 to 60 cm, it ranged from 0.13 to 0.15 c mol kg^{-1}. Magnesium content was higher at the upper layer of the soil ranging from 1.03 to 2.51 (c mol kg^{-1}) in the upper 30 cm while at 30 to 50 cm it ranged from 0.94 to 1.06 c mol kg^{-1}.

Compared with site used in 2011, the OM was higher ranging from 2.05 to 2.42% in the upper 30 cm. OM decreases down the soil profile being 0.64% at 50 cm. Phosphorus content decreases down the soil profile. The soil pH ranged from 5.14 to 5.38 within a depth of 60 cm. The available nutrients in the soils in the two seasons are adequate for the cultivation of the crop without the addition of artificial fertilizer for improvement of soil fertility.

7.3   Moisture content and temperature of the soil

The distribution of the soil moisture and temperature are described in the sections that follow.

7.3.1 Distribution of moisture in the soil

The soil moisture contents were characterized by fluctuations along the crop cycle and this can be attributed to variability in rainfall during the growing seasons (Table 7.1). This fluctuation was more pronounced among the treatments in 2011 (Figure 7.1) with lower seasonal rainfall compared with the year 2012. The upper 30 cm of the soil where more that 95% of the active roots are located is much wetter than the lower 20 cm in 2011. The average soil moisture contents among the treatments in the upper 10 to 30 cm and lower 30 to 60 cm in 2011 are 0.20 and 0.14 $m^3 m^{-3}$ respectively after establishment of the crop in 2011 while in 2012, the corresponding soil moisture contents are 0.24 and 0.25 $m^3 m^{-3}$ respectively (Figure 7.1).

Table 7.2. Physical and chemical properties of the soil at the experimental fields during the rainy seasons

Property/depth (cm)	2011						2012					
	00-10	10-20	20-30	30-40	40-50	50-60	00-10	10-20	20-30	30-40	40-50	50-60
Sand (%)	72	72	75	71	72	73	69	63	59	54	51	52
Clay (%)	22	14	15	17	20	20	15	22	25	31	34	33
Silt (%)	06	14	10	12	08	7	16	16	16	15	15	15
Textural class	Sandy clay loam	Sandy loam	Sandy loam	Sandy loam	Sandy loam	Sandy loam	Sandy loam	Sandy clay loam	Sandy clay loam	Sandy clay loam	Sandy clay loam	Sandy clay loam
Bulk density (g cm^{-3})	1.45	1.53	1.60	1.58	1.61	1.62	1.45	1.56	1.61	1.64	1.65	1.66
Organic matter (%)	1.95	2.04	1.57	1.05	0.64	0.64	2.42	2.33	2.05	1.95	1.88	1.84
H$_2$CO$_3$ (ppm)	42	35	58	47	40	43	65	69	73	99	72	70
Phosphorus (ppm)	15.8	16.0	15.8	16.8	15.8	15.7	23	24.3	23.3	22.6	19.2	19.1
Na$^+$ (c mol kg^{-1})	0.12	0.12	0.12	0.12	0.12	0.12	0.12	0.12	0.13	0.13	0.13	0.13
K$^+$ (c mol kg^{-1})	0.17	0.21	0.15	0.15	0.13	0.14	0.24	0.23	0.21	0.20	0.21	0.21
Ca^{2+} (c mol kg^{-1})	2.35	2.30	1.67	1.28	1.48	1.45	1.34	1.91	1.90	1.87	1.37	1.37
Mg^{2+} (c mol kg^{-1})	2.51	2.51	1.03	0.94	1.06	1.05	1.24	1.11	1.20	1.55	1.55	1.56
EC (dSm^{-1})	0.20	0.30	0.00	0.00	0.03	0.02	0.10	0.10	0.20	0.00	0.00	0.01
pH	6.4	6.5	6.3	6.4	6.3	6.30	5.14	5.25	5.34	5.45	5.38	5.25
S	17.6	18.7	17.8	17.4	18.2	18.5	17.2	19.1	17.9	17.2	25.4	25.1
Cu^{2+}	5.68	5.84	5.57	5.17	4.31	4.28	5.87	5.88	5.58	5.16	4.32	4.31
Fe^{2+}	16.7	16.2	15.3	14.2	13.6	13.7	16.4	15.9	15.1	14.0	13.7	13.2

In 2011, the average moisture content in the upper 30 cm in the ML was 17% higher than that of the NC, while the lower 30 to 60 in TRBD was 22% higher than that of NC at 42 DAP (Anthesis). Similarly, in the upper 30 cm, average soil moisture content under TRML and TR were 16.4 and 15% respectively higher than that of NC. The higher average moisture content in the upper soil layer under the treatments where mulch was applied was due to the protective layer provided by the mulch materials against evaporation of water from the soil. This is similar to the findings of (Munn, 1992) in Wheat plantation and (Smith, and Rakow, 1992) in woody landscape planting and (Kraus, 1998) in the cultivation of desert willow. Different levels and type of mulches also increased moisture content of soil and water status of soil in tea production in Kenya (Othieno, 1980). The CV of the soil moisture content at the upper 30 cm among the treatments in 2011 at 63 DAP was 0.08% while at 98 DAP it was 0.06% (Figures 7.1 and 7.2). The CV in 2012 at 60 and 102 DAP among the treatments were 0.03 and 0.07%. Impermeable layer of soil was present at a depth beyond 60 cm at the experimental field in the first season.

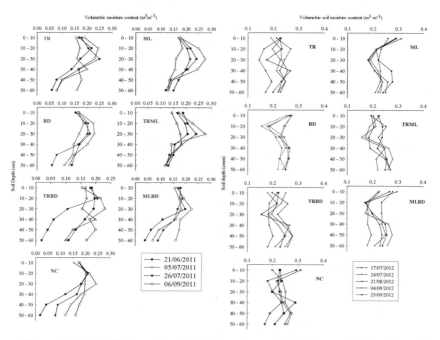

Figure 7.1. Variability in volumetric moisture content in the upper 60 cm of the soil profiles under rainfed conditions

7.3.2 Temperature in the soil

The average temperature in the upper 30 cm of the soil profile for control treatment was 29.3 °C, but 27.1, 26.6 and 26.4 for TR, ML and TRML respectively shortly after establishment in year 2011(Figure 7.3). The average temperatures in the upper 30 cm in the Control treatment (NC) were 7.6, 9.3, and 9.6% higher than in TR, ML and TRML (Appendix D). The reduction in the soil temperature under ML could be attributed to the presence of plant residues, which provided a protective layer and shielded the soil surface from direct impact of SR. Similarly, in the lower 20 cm in the soil profile, the average temperatures are 28.7 °C for NC while it was 26.6, 26.8 and 26.5 °C for TR, ML and TRML respectively. The CV of the soil temperature in among the treatments in the upper 30 and lower 30 cm was 4.0%. During pod initiation however, there was a reduction in the soil temperatures (Figure 7.3) at both the upper and the lower profiles with CV of 1 and 5%, respectively. During the pod initiation when the soil temperature was lower, average global SR was 125 W m^{-2} and lower compared to the preceding months of May and June whose global SR were 172 and 156 W m^{-2} respectively (Appendix C). Lowest temperature was 24.9 °C in the upper and lower 30 cm of the soil profiles in August.

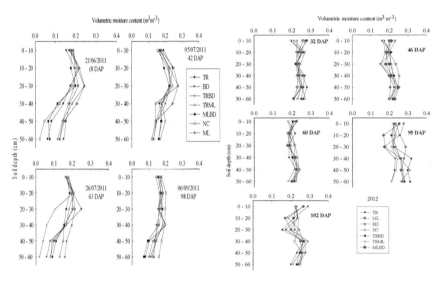

Figure 7.2. Variability in volumetric moisture content among the treatments during establishment, flowering, pod initiation, seed filling and near maturity under rainfed conditions. Each dot represents the mean of the soil moisture from two replicates

Higher temperature at the upper soil profile in 2012 is similar to that of 2011 at the development stage of the crop. The average soil temperature in the upper 30 cm for NC was 26.9 °C, while in the ML and BD they were 26.5 and 27.0 °C respectively (Appendix D). In the lower 20 cm of the soil profile, the average temperature was 26.9 °C for BD plot but 26.4 and 26.9 °C for NC and TRBD. The coefficient of variation in the soil temperature was 1% at both layers in 32 DAP which resulted in no significant difference in the average soil temperature in the upper 30 and lower 20 cm of the soil profiles among the treatments. However, at the commencement of maturity, the presence of mulch resulted in higher temperature compared with NC. High leaf area at the mid season could also be responsible for the reduction in the soil temperature because higher canopy cover intercepted greater proportions of the incident SR and reduced the soil temperature. The Soil bund had the lowest temperature of 22.3°C in the upper 10 to 20 cm of the soil in August. Lowest global SR were 117 and 101 W m^{-2} respectively for the month of August in both seasons (Appendix C). Low global SR coupled with the use of mulch as protective layer was responsible for reductions in the soil temperatures in the treatments in July and August in both seasons (Figure 7.3).

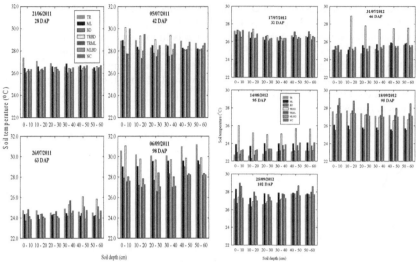

Figure 7.3. Average temperatures in the upper 60 cm of the soil profiles under rainfed conditions. Each bar represents the mean of the soil temperatures from two replicates

7.4   Seasonal soil water storage (SWS)

Figure 7.4 shows the SWS at five growth stages namely the vegetative VE-V2 and reproductive which are flowering - R1; pod initiation- R3, seed filling R6 and maturity. Rainfall depths for August and September in the years 2011 and 2012 constituted 32 and 41%, respectively of the seasonal rainfall. At the vegetative stage, 26 DAP in 2011, SWS of 101±7.07 mm for TRML was significantly higher than those of TR and ML at α = 5% level. TRML, TR and ML did not produce significant difference in the SWS until maturity of the crop in the first season. However, SWS in TRML was 25, 24 and 16% greater than those of NC, ML and BD respectively at the initial stage in June. In 2012 there was no significant difference in SWS in July (28 DAP) except for the BD plot that had the highest water storage of 127±4.95 mm and was higher than MLBD, ML and NC by 12.6, 9.45 and 8.66% respectively. The SWS reduced from 101±7.1 mm during vegetative stage in June to 95.0±4.24 mm during seed filling stage in August, 2011. This was due to a commonly called August break in rainfall in the study area and other places in the South Western part of Nigeria. During seed filling stage, in August (97 DAP), SWS for TRBD was higher than that of BD, TR and NC by 21, 17.4 and 20.4% respectively but there was no significant difference (p < 0.05) in the SWS among the treatments. However, there were significant differences in the SWS among the treatments in the second season. The ML had the highest water storage of 151±12.1 mm during the seed filling stage in September (98 DAP) and greater than SWSs in NC, MLBD and TR by 19, 24 and 26% respectively, despite a break in rainfall in August. The peak rainfall in September compensated for the drought in August in the second season (Table 7.1). Seasonal SWSs were in the order of NC < BD < MLBD < ML< TR< TRBD < TRML in the first season while in the second season, it was in the order of TR< TRML< MLBD < NC< BD < ML < TRBD. For instance, the seasonal SWS for BD, MLBD, ML, TR, TRBD and TRML were 6.44, 6.65, 8.54, 9.96, 10.6 and 14.5% higher than that of NC in the first season. However, in the second season, the seasonal

SWS for BD, ML and TRBD were 0.53, 1.92 and 2.77% respectively higher than that of NC. The seasonal SWS in 2012 were higher in all the treatments than those of the first season and this was due to differences in the seasonal rainfall and changes in the soil conditions at the experimental sites. TRBD had the highest seasonal SWS of 578 mm in the second season while in the first season it was 476 mm for TRML (Appendix E). Experimental field in the second season (Table 7.2) contains higher percentage of clay and therefore has greater water retaining capacity than sandy soil at the experimental field in 2011(Kirkham, 2004). This in addition to the conservation practices accounted for the higher seasonal SWS in the second season in the treatments.

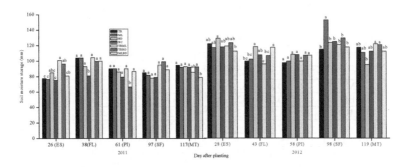

Figure 7.4. Soil water storage at five growth stages of the plant during rainfed conditions. Each bar represents average soil water storage from two replicates. Bars with the same letter indicate that the soil water storages are not significantly different at p < 0.05 by using Duncan multiple comparison of means

7.5   Biometric measurements and analysis of the plant canopy

Results of the biometric data namely the plant heights, number of leaf, LAI are in the sections that follow.

7.5.1   Plant height and number of leaves

Table 7.3 shows the average plant heights at five growth stages. There was variability in the plant heights among the treatments under the rainfed conditions. At flowering stage in 2011, there was no significant difference in the average plant heights in TRML, TRBD and MLBD despite variability in the SWS unlike in 2012. At maturity, NC had the lowest seasonal SWS of 407 mm and peak plant height of 67.8±1.3 cm. The peak height was higher than those of MLBD, TRBD and NC by 22.1, 14.0 and 21.7% respectively. In 2012, the application of mulch and creation of side bund to conserve soil water in TRML, TRBD and MLBD did not produce any significant difference in the plant heights compared with the NC during the pod initiation stage. At this stage, average plant height for NC was significantly higher than those of other treatments. The highest average height of 80.3±0.6 cm at maturity in 2012 was obtained for MLBD whose seasonal SWS was 558 mm. The peak height for MLBD was higher than TRBD, TRML and NC by 24.9, 7.5 and 3.7% respectively. Plant heights at harvest were significantly higher in the treatments where water was conserved. This could be attributed to higher SWS that favoured shoot growth and foliar development of the plant in the second season. Pooled data of plant heights at harvest and seasonal SWS for the two seasons were significantly correlated ($r^2 = 0.74$). The t-test shows that average plant height in the second rainy season was significantly higher than that of the first season at $\alpha = 0.05$. The plant heights in the two seasons compared favourably with the results of similar experiment at Southern Guinea savannah in Ilorin on TGX 1448 2^E where the seasonal average height and rainfall were 57.8 cm and 610 mm respectively (Aduloju et al., 2009). There was no increment in the plant height after seed filling in the two seasons in all the treatments.

Table 7.3. Average plant heights at five different growth stages under rainfed conditions

		Vegetative stage	Reproductive stage			
		Initial (VE-V2)	Flowering (R1)	Pod initiation (R3)	Seed filling (R6)	Maturity (R8)
2011	TR	18.4 ± 0.4^c	0.2 ± 2.0^b	52.0 ± 0.1^b	52.3 ± 0.6^{cd}	52.3 ± 0.6^{cd}
	ML	19.6 ± 1.1^{ab}	33.7 ± 0.9^a	50.5 ± 5.3^b	52.0 ± 1.7^d	52.1 ± 1.7^{de}
	BD	18.9 ± 0.4^{bc}	32.7 ± 1.8^{ab}	51.8 ± 0.5^b	51.3 ± 1.0^e	51.3 ± 1.0^e
	TRML	20.1 ± 0.8^a	32.7 ± 1.2^{ab}	51.3 ± 6.1^b	53.1 ± 0.6^c	53.1 ± 0.1^c
	TRBD	18.2 ± 0.2^c	32.6 ± 1.3^{ab}	56.8 ± 1.7^{ab}	58.3 ± 0.6^b	58.3 ± 0.5^b
	MLBD	19.1 ± 0.2^{bc}	33.3 ± 2.0^{ab}	52.8 ± 4.0^b	52.8 ± 0.6^a	52.8 ± 0.3^{cd}
	NC	20.3 ± 0.7^a	34.7 ± 3.6^a	61.5 ± 8.5^a	67.8 ± 0.6^a	67.8 ± 1.3^a
	Average	19.2 ± 2.2	34.8 ± 5.6	53.8 ± 5.6	55.0 ± 5.6	55.4 ± 5.6
2012	TR	24.3 ± 0.6^a	43.0 ± 0.4^a	50.7 ± 11.6^a	70.3 ± 0.6^e	72.0 ± 2.0^d
	ML	18.3 ± 0.6^c	32.3 ± 0.8^{cd}	47.7 ± 0.5^a	72.0 ± 1.8^d	72.0 ± 0.6^d
	BD	18.0 ± 0.1^c	35.0 ± 0.3^b	53.0 ± 0.5^a	76.0 ± 1.1^b	76.0 ± 0.6^c
	TRML	16.6 ± 0.6^d	31.8 ± 0.4^d	50.0 ± 1.0^a	74.3 ± 0.6^c	74.3 ± 0.6^c
	TRBD	16.3 ± 0.5^d	29.7 ± 0.6^e	49.7 ± 0.6^a	60.3 ± 0.6^f	60.3 ± 0.6^e
	MLBD	20.3 ± 0.8^b	35.0 ± 0.2^b	55.7 ± 1.8^a	80.3 ± 0.6^a	80.3 ± 0.7^a
	NC	17.7 ± 1.0^c	32.7 ± 0.7^c	50.0 ± 1.0^a	77.3 ± 0.6^b	77.3 ± 0.6^b
	Average	18.8 ± 2.1	34.2 ± 5.4	51.1 ± 5.1	72.9 ± 5.6	73.0 ± 5.3

Values are mean ± SD from four replicates. Average plant heights with the same letter are not significantly different at 5% (p > 0.05) level based on Duncan comparison of means.

The peak number of leaves was obtained during seed filling stage in each season in all the treatment including NC. The peak number of leaf was recorded for NC in the first season (Table 7.4).

7.5.2 Extinction coefficient, Leaf area index and fraction of Intercepted Photosyntheticaly Active Radiation

Leaf extinction coefficient (λ)

Seasonal λ ranges from 0.40 for TRML to 0.48 for ML, BD and TRBD, with an average of 0.46±0.04 in the first season (Figure 7.5). However, in 2012, it ranged from 0.47 for MLBD to 0.55 for ML, which produced a seasonal average of 0.51±0.03 (Figures 7.5). The t-test shows that means of λ s in the two years are not significantly different (p > 0.05, $\alpha = 0.05$). Average seasonal λ vary even for a single cultivar of Soybeans. Lower seasonal λs for TR, TRML in the first season, TR and MLBD in the second season show that the leaves maintained upright positions under field conditions (Mavi and Tupper, 2004) which ensured more uniform distribution of photosynthesis through the canopy (Hammer et al., 2009). Therefore, a lower value of λ when combined with higher LAIs is a virtue in the current cultivar of Soybeans. The λ obtained in the study for Soybeans compares well with those found in literature. Kiniry et al. (1992) reported λ of 0.45 for

Soybean. Flenet (1996) reported an average of 0.52, 0.43 and 0.32 for row spacing of respectively 0.35, 0.66 and 1.00 m in Texas.

Table 7.4. Average number of leaves at five growth stages under rainfed conditions

Year	Vegetative stage		Reproductive stage			
	Treatment	Initial (VE-V2)	Flowering (R1)	Pod initiation (R3)	Seed filling (R6)	Maturity (R8)
2011	TR	20 ± 0.6^{bc}	43 ± 9.0^{c}	241 ± 1.6^{c}	248 ± 0.12^{f}	16 ± 1.0^{f}
	ML	19 ± 1.2^{c}	48 ± 2.3^{bc}	220 ± 22.6^{c}	226 ± 0.82^{g}	14 ± 0.8^{g}
	BD	21 ± 0.1^{b}	$45\pm3.4b^{c}$	237 ± 56.2^{c}	252 ± 0.0^{d}	16 ± 0.6^{d}
	TRML	22 ± 0.1^{a}	59 ± 7.7^{a}	258 ± 36.5^{bc}	257 ± 0.8^{c}	16 ± 0.8^{c}
	TRBD	20 ± 1.0^{bc}	61 ± 5.6^{a}	308 ± 5.9^{ab}	337 ± 1.1^{b}	22 ± 1.0^{b}
	MLBD	21 ± 1.0^{b}	50 ± 7.7^{bc}	243 ± 51.9^{c}	250 ± 0.6^{e}	16 ± 1.0^{e}
	NC	22 ± 0.10^{a}	54 ± 8.1^{ab}	330 ± 24.5^{a}	343 ± 0.96^{a}	22 ± 1.0^{a}
	Average	21 ± 1.16	51 ± 8.51	262 ± 49.1	273 ± 0.85	18 ± 0.79
2012	TR	20 ± 0.6^{a}	56 ± 0.6^{d}	139 ± 32.1^{c}	265 ± 0.6^{e}	15 ± 0.6^{e}
	ML	20 ± 1.0^{a}	64 ± 0.7^{b}	141 ± 0.5^{c}	311 ± 0.6^{c}	26 ± 1.1^{c}
	BD	20 ± 0.6^{a}	60 ± 0.5^{c}	174 ± 1.0^{ab}	310 ± 1.0^{c}	19 ± 0.6^{d}
	TRML	18 ± 1.7^{b}	54 ± 0.6^{e}	154 ± 0.6^{bc}	347 ± 2.7^{a}	16 ± 0.6^{e}
	TRBD	17 ± 0.6^{b}	55 ± 1.0^{e}	160 ± 0.6^{abc}	260 ± 1.1^{f}	79 ± 0.6^{a}
	MLBD	21 ± 0.6^{a}	72 ± 0.6^{a}	178 ± 1.5^{a}	322 ± 1.2^{b}	50 ± 0.8^{b}
	NC	21 ± 0.6^{a}	61 ± 1.1^{c}	154 ± 0.6^{bc}	290 ± 1.0^{d}	15 ± 0.5^{e}
	Average	19 ± 0.63	60 ± 0.72	157 ± 0.26	300 ± 1.14	31 ± 0.64

Values are mean ± SD from four replicates. Average numbers of leaves with the same letter are not significantly different at 5% (p < 0.05) level based on Duncan comparison of means.

LAI

The LAIs of the plants in the two seasons follow a similar trend from the initial stage (establishment) and gradual falling during seed maturity (R7) at the late stage (Table 7.5). Daily trend of the LAIs along the growing season is in Figure 7.6. Results of the statistical analysis of the variability in the LAIs are shown in Appendix F. At 26 DAP (DOY 170) in (VE-V2) growth stage in the first season, the LAIs were 0.49, 0.58, 0.65 and 0.54 m^2 m^{-2} for TR, ML, BD and MLBD, respectively (Figures 7.8). Average LAI from four replicates of all the treatments was 0.55 ± 0.14 m^2 m^{-2} and there was no significant difference (p > 0.05) in the LAIs at the end of the early stage of the growth in the first season. During flowering (R1 growth stage), the LAIs in MLBD and TRML were 4.60 and 1.19% respectively higher than that of NC. No significant differences were found in the LAIs at flowering stage among the seven treatments. However, the LAIs of TRBD, TR, ML and BD were 2.47, 13.7, 13.7 and 23.8% lower than that of NC at this stage in the first season. The average LAI in the second season among the treatments was 1.01 ± 0.09 m^2 m^{-2} and higher than the average in the first season 0.78 ± 0.18 m^2 m^{-2} after the commencement of flowering and this can be attributed to higher water storages (Figure 7.4) in the second season. At maturity, the LAIs in MLBD, TRML, TRBD and BD were 7.73, 10.2, 14.4 and 20.5% higher than that of NC while the LAIs of TR and ML were 9.2 and 36.9% lower than that of NC. Despite the differences in the values of LAIs in the treatments compared with NC at maturity, there were no significant differences (p < 0.05) in the LAIs of BD, TRML and TRBD whose seasonal SWS were 435, 476 and 455 mm respectively. Significant differences at 5% level of comparison were only found in the LAIs of BD, NC, MLBD and TR whose seasonal SWS were 435, 407, 436 and 452 mm respectively.

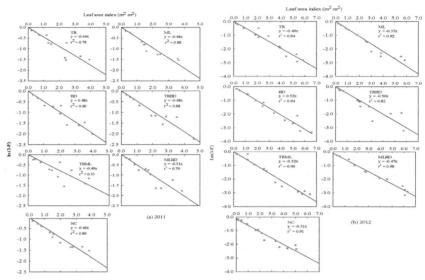

Figure 7.5. Relationship between natural log of the intercepted PARs and LAIs in 2011and 2012. The slopes of the regression lines are the extinction coefficients of the crop

In the second season at flowering stage, the LAIs for TRML, BD and MLBD were 21.8, 24.4 and 32.1% respectively higher than that of NC while those of TR, ML and TRBD were 6.90, 14.8 and 32.9% respectively lower than that of NC. Significant differences were found between the LAIs for TR, MLBD, ML and TRBD. After maturity of the seed in the two seasons, there was a reduction in the LAIs. The reduction in the LAIs at maturity was due to a fall in the photosynthetic capacity of the leaves after the plants had completed their life cycle as reported in (Burton, 1997). This occurred because the Nitrogen in the leaves has been exported to the seed and the rate of nitrogen fixation had reduced substantially. At maturity, the LAIs of MLBD, ML, TRML, BD and R were 17.2, 48.9, 61.6, 75.1 and 77.5% respectively higher than that of NC whereas, the LAIs of TRBD was lower than that of NC by 37.1%. The seasonal SWS were 543, 548, 558 and 562 mm for TR, TRML, MLBD and NC respectively. Others were 565, 573 and 578 mm for BD, ML and TRBD respectively. The LAIs of the six water conservation practices and the conventional practice were significantly different from one another at maturity in the second season (Table 7.5).

There were no significant differences in the LAIs in the first season until during pod initiation (Table 7.5). The peak LAIs were 4.78 ± 0.50 m^2 m^{-2} (DOY 241) and 6.16 ± 0.04 m^2 m^{-2} (DOY 265) for TRBD and MLBD, respectively whose seasonal SWS were 455 and 558 mm in both seasons when the pod had been completely filled with seed (R6 growth stage). This could be attributed to higher SWS in the Soil bunded plots in the two seasons (Figure 7.4).

Table 7.5. Leaf area indices at different growth stages of the crop under rainfed conditions

		Vegetative stage	Reproductive stage			
		Initial (VE-V2)	Flowering (R1)	Pod initiation (R3)	Seed filling (R6)	Maturity (R8)
2011	TR	0.49±0.06^a	0.73±0.18^a	2.35±0.61^a	3.97±0.17^{ab}	1.22±0.01^e
	BD	0.65±0.16^a	0.68±0.09^a	2.01±0.35^{abc}	4.13±0.35^{ab}	2.10±0.10^a
	ML	0.58±0.06^a	0.73±0.09^a	1.41±0.14^c	3.65±0.34^{ab}	1.53±0.14^d
	TRML	0.46±0.14^a	0.84±0.28^a	1.67±0.38^{bc}	4.56±1.52^{ab}	1.86±0.23^{abc}
	TRBD	0.59±0.18^a	0.81±0.09^a	2.24±0.31^{ab}	4.78±0.50^a	1.95±0.01^{ab}
	MLBD	0.54±0.15^a	0.87±0.29^a	1.56±0.20^c	2.83±1.87^b	1.81±0.12^{bc}
	NC	0.58±0.15^a	0.83±0.11^a	1.89±0.42^{abc}	4.08±1.60^{ab}	1.67±0.31^{cd}
	Average	0.55±0.14	0.78±0.18	1.87±0.37	4.00±1.12	1.73±0.17
2012	TR	0.23±0.01^d	0.87±0.01^a	2.12±0.01^a	5.33±0.02^d	2.13±0.01^a
	BD	0.27±0.01^c	1.23±0.01^{ab}	2.49±0.01^c	5.37±0.01^d	1.93±0.01^b
	ML	0.27±0.01^c	0.81±0.57^c	2.57±0.01^b	5.54±0.02^c	0.94±0.02^d
	TRML	0.29±0.01^b	1.19±0.01^{ab}	2.82±0.01^a	6.03±0.02^b	1.25±0.02^c
	TRBD	0.26±0.01^c	0.70±0.01^c	1.95±0.01^f	4.70±0.04^f	0.35±0.01^g
	MLBD	0.35±0.01^a	1.37±0.01^a	2.82±0.02^a	6.16±0.04^a	0.58±0.01^e
	NC	0.19±0.01^e	0.93±0.01^{bc}	2.20±0.01^d	5.06±0.04^e	0.48±0.01^f
	Average	0.27±0.05	1.01±0.29	2.42±0.32	5.46±0.49	1.10±0.66

Values are mean ± SD from four replicates. Means LAI with the same letter are not significantly different at 5% (p > 0.05) level based on Duncan comparison of means.

Lower LAIs in the first season compared to the second season even during seed filling stage could be attributed to the response of the crop to water stress caused by reduction in rainfall in August. There was reduction in stomata conductance, which led to interruption in CO_2 assimilation. Plants could die if the stress continues for long because the water reserves could have been exhausted (Sinclair, 2000). The reduction in LAIs in the first season was a strategy by the crop to maintain soil water uptake at a satisfactory level. The LAIs were significantly higher in the treatments where the fields were managed to conserve more water in the root zone.

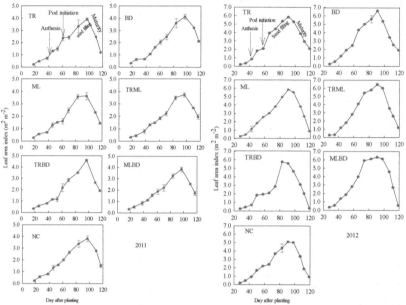

Figure 7.6. Changes in the leaf area indices of Soybeans along phenologic stages in the rainy season. Each dot represents the mean with standard error of four replicates in the treatments

fIPAR and LAI

The fIPARs at the early stage of growth in the two seasons were similar (Table 7.6). Output of statistical analysis of the fIPAR is shown in Appendix G. At the initial stage, that is V2 growth stage in the 2011 season, the fIPAR ranges from 0.21 to 0.27. Field management by using Mulches and Soil bund did not have significant effects on the fIPAR at the initial stage, as the values of fIPAR were nearly the same compared with NC. The average fIPAR in the initial stage in the first season was 0.25±0.01 compared with 0.17±0.01 in the second season. At flowering stage in the first season, the MLBD had the peak fIPAR while in the second season; the ML had the peak fIPAR. This is attributed to higher LAI under MLBD (Table 7.6) in both seasons. Highest interception of PAR was recorded during the seed filling stage (R6) of growth in both seasons. During this stage, Soil bunded plots had higher fIPAR, 0.91±0.01 (DOY 241) for TRBD and 0.95±0.04 (DOY 265) for MLBD in the first and second seasons, respectively (Table 7.7). At 96 DAP (seed filling stage) in the first season, fIPAR of 0.91±0.01 in September (DOY 241) for TRBD was 12.2% higher than those of ML and NC; 13.3% higher than TR and 3.3% higher than BD. However, in 2012, peak fIPAR was

0.95±0.04 for TRML and MLBD was 8.42% higher than TRBD but 7.3% higher than NC and BD. The fIPAR of the crop at this stage could have reduced if there was significant water stress and this could lead to reduction in the yield components (Sinclair, 2000). Average fIPAR reduced to 0.54±0.03 (DOY 261) and 0.35±0.02 (DOY 287) in the first and second seasons, respectively at maturity (R8 growth stage). The progressive reduction in the fIPARs after seed maturity is attributed to senescence and reduction in the capacity of the plant at R7-R8 growth stages to utilize radiation for biomass formation because the plant had completed its life cycle. There were strong, positive and significant correlations (p < 0.05) between fIPARs and the LAIs ($0.70 \leq r^2 \leq 0.99$) in the first season and ($0.93 \leq r^2 \geq 0.99$) in the second season by using an exponential model (Figure 7.7). This indicates that the LAIs accounts for between 70 to 99% of the variability in fIPARs of the crop in first and second seasons respectively and can be used to accurately predict CC of the crop in the study area at any time of the crop cycle. Good correlation between LAIs and fIPAR in this study is similar to those found in giant reeds and sweet Sorghum (Ceotto, 2013). Higher air temperatures in the first season (Table 7.1) caused the curling of the leaves, reduction in LAIs and resulted in lower fIPAR compared to the same LAIs in the second season.

Table 7.6. Average fIPAR at different growth stages of Soybean during the rainy seasons

		Vegetative stage		Reproductive stage			
		Initial (VE-V2) (26 DAP)	Flowering (R1) (38 DAP)	Pod initiation (R3) (61 DAP)	Seed filling (R6) (96 DAP)	Maturity (R8) 117 DAP
2011	TR	0.27±0.01[a]	0.31±0.01[c]	0.77±0.01[a]	0.78±0.01[e]	0.31±0.01[d]
	BD	0.26±0.01[a]	0.29±0.01[d]	0.65±0.01[c]	0.88±0.01[b]	0.51±0.01[c]
	ML	0.26±0.01[a]	0.29±0.01[d]	0.69±0.01[b]	0.79±0.01[ed]	0.56±0.01[bc]
	TRML	0.22±0.02[b]	0.34±0.01[b]	0.70±0.02[b]	0.80±0.01[d]	0.53±0.03[bc]
	TRBD	0.26±0.01[a]	0.34±0.01[b]	0.70±0.01[b]	0.91±0.01[a]	0.68±0.01[a]
	MLBD	0.21±0.01[b]	0.40±0.01[a]	0.63±0.01[d]	0.83±0.01[c]	0.55±0.02[bc]
	NC	0.26±0.01[a]	0.35±0.01[b]	0.65±0.01[c]	0.79±0.01[ed]	0.61±0.01[ab]
	Average	0.25±0.01	0.33±0.01	0.69±0.01	0.83±0.01	0.54±0.03
		28 DAP	43 DAP	58 DAP	98 DAP	119 DAP
2012	TR	0.15±0.01[c]	0.41±0.01[d]	0.79±0.01[a]	0.92±0.02[b]	0.42±0.02[b]
	BD	0.17±0.01[b]	0.42±0.01[cd]	0.76±0.01[a]	0.88±0.01[c]	0.65±0.24[a]
	ML	0.17±0.01[b]	0.47±0.01[a]	0.76±0.01[a]	0.93±0.02[ab]	0.43±0.02[b]
	TRML	0.16±0.01[b]	0.44±0.01[b]	0.76±0.01[a]	0.95±0.02[a]	0.18±0.01[e]
	TRBD	0.13±0.01[d]	0.30±0.01[e]	0.67±0.01[a]	0.87±0.04[c]	0.24±0.01[d]
	MLBD	0.24±0.01[a]	0.43±0.01[cb]	0.79±0.02[a]	0.95±0.04[a]	0.28±0.01[c]
	NC	0.15±0.01[c]	0.42±0.01[cd]	0.71±0.01[a]	0.88±0.04[c]	0.28±0.01[c]
	Average	0.17±0.01	0.41±0.09	0.75±0.01	0.91±0.03	0.35±0.02

Values are mean ± SD from four replicates. Means of the fIPAR with the same letter are not significantly different at 5% (p > 0.05) level based on Duncan comparison of means.

Cumulative IPAR (TIPAR) and DAB

According to Monteith (1981) TIPAR over the crop season is an important determinant of production of any crop. The TIPARs throughout the growing season ranged from 189 MJ m^{-2} for TR to 218 MJ m^{-2} for BD in the first season while in the second seasons, it ranged from 248 MJ m^{-2} for TRBD to 284 MJ m^{-2} for MLBD (Table 7.7).

Table 7.7. Comparison of the RUEs determined by using PARs from global solar radiation and instantaneous measurements of PARs during the rainy seasons

Year (1)	Treatment label (2)	RUE by using PAR from the daily global solar radiation (g MJ^{-1} IPAR) (3)	RUE by using Instantaneous measurement of PAR (g MJ^{-1} IPAR) (4)	Absolute difference Δ (3-4) (5)	Cumulative IPAR from solar radiation (MJ m^{-2}) (6)
2011	TR	1.18	0.80	0.38	189
	BD	1.98	1.12	0.86	218
	ML	1.92	1.29	0.63	205
	TRML	1.94	1.30	0.64	209
	TRBD	1.95	1.65	0.30	216
	MLBD	1.92	1.07	0.85	209
	NC	1.94	1.08	0.86	211
	Average	1.83	1.18	0.65	208
2012	TR	1.45	0.94	0.51	275
	BD	1.81	1.24	0.57	284
	ML	1.92	1.24	0.68	273
	TRML	1.73	1.11	0.62	279
	TRBD	1.43	0.96	0.47	248
	MLBD	1.82	1.08	0.74	284
	NC	1.89	1.06	0.83	256
	Average	1.72	1.09	0.63	271

The RUE was determined from the linear regression of the average daily biomass and TIPAR from four replicates of each treatment.

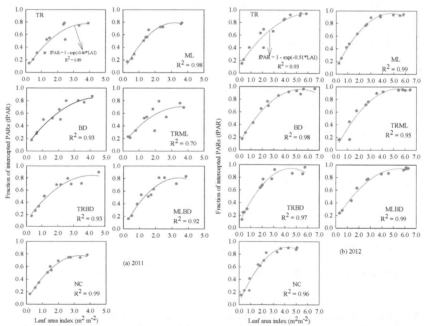

Figure 7.7. Leaf area indices and fIPAR at extinction coefficients of 0.46 and 0.51 during the rainy seasons. Each dot represents the mean of LAIs from four replicates

In the two seasons, the trend in the biomass accumulation was similar. The DAB increased from initial stage and got to the peak at mid season before descending at the late stage of the plant cycle (Figure 7.8). At the initial stage (V2 growth stage), 21 DAP in June 2011 the average cumulative DAB among the treatments ranged between 3.00 g m^{-2} and 5.18 g m^{-2} in the first season while in the second season it ranged between 4.50 g m^{-2} and 7.70 g m^{-2} 28 DAP in July. DAB for BD was 57% higher than TRML, TRBD and TR but 43% higher than the ML in 2011. The DAG biomass in BD plot was 25% higher than ML and NC but 37.5% higher than TRBD and TRML in 2012. The peak DAB of 348 g m^{-2} in TRBD in the first season (102 DAP in September) was 67 and 66.1% higher than the DAB for NC and MLBD respectively. However, it was higher than BD and TR by 46.8%. Similarly, in 2012, 105 DAP in September peak DAB of 437 g m^{-2} for BD was 34.2 and 36.9% higher than the DAB for NC and TRML respectively but 50.6% higher than DAB for TRBD. Peak DAB were recorded in September in the two seasons shortly after the commencement of maturity at R7-R8 growth stage. DAB in 2011were lower compared with corresponding biomasses at the same stages in the 2012 (Figure 7.8).

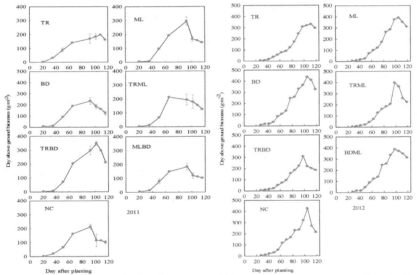

Figure 7.8. Daily cumulative above ground biomass from emergence until maturity under rainfed conditions. Each dot represents the mean and standard error for four replicates

Linear model fitted to the TIPARs and DAB at harvest in both seasons shows that they are significantly and strongly related (DAB (t ha^{-1}) = 0.02*IPAR - 2.88) (r^2 = 0.76, p < 0.05). This indicates that the linear models account for 76% of the variability in dry matter at harvest in relation to TIPAR. Similar high correlations were observed for, for Soybean and Maize (Singer et al., 2011), for Pigeon pea (Saha et al., 2012) and mustard (Pradhan et al., 2014). This means that increase IPAR is central and is a key driving force for accumulation of biomass in the crops (Latiri-Souki et al., 1998).

The variability in seasonal SWS also contributed to the differences in the biomass at harvest in the two seasons (Figure 7.4). For instance, TRML that had the highest SWS of 476 mm produced cumulative biomass of 125 g m^{-2} while NC whose seasonal SWS was 407 mm produced 101 g m^{-2} at harvest. The peak biomass at harvest of 211 g m^{-2} for TRBD in the first season was higher than those of ML, TRML and NC by 32.7, 40.8 and 52.1% respectively. Similarly in 2012, the seasonal SWS of 565 mm produced dry biomass of 325 g m^{-2} for BD. The dry biomass was higher than those of ML, TRML and NC by 4.3, 31.1 and 34.8% respectively. Pooled over the years, SWS and DAB at harvest are significantly correlated ($r^2 = 0.79$, $p < 0.05$). This indicates that 79% of the variability in dry biomass at harvest can be explained by SWS within the root zone. The t-test shows that the seasonal average biomass at harvest in the second season was significantly higher than that of the first season. The treatments where water was conserved in the seasons had higher cumulative DAB compared with the conventional practice. Among all the treatments, the Soil bunded (BD) plot produced the highest biomass at harvest. Higher rainfall resulted in higher SWS, on the field in the second season (Table 7.3) and that the second season was more humid than and favoured higher biomass accumulation than the first.

7.6   Radiation Use Efficiency

The PAR constituted 41% of the SR (PAR = 0.41SR, $r^2 = 0.99$). This value is close and compares well with and 0.45 in Ilorin; Nigeria (Udor and Aro, 1999), 0.44 in semi Arid environment in Australia (O' Cornell et al., 2004); about 0.40 in Monteith and Unsworth (2013); 0.48 recommended for converting SR to PAR (Soltani and Sinclair, 2012) and 0.47 in Howell et al. (1983). Strong linear relationships were found between the TIPARs and DAB until seed filling (R6 growth stage) in the two seasons in all the treatments (Figures 7.10). Coefficients of determination were $0.83 \le r^2 \le 0.95$ in 2011 and $0.81 \le r^2 \le 0.96$ in 2012. The linear relationships stopped at 70 DAP for TRML; 96 DAP for ML and MLBD but 110 DAP for TRBD and BD in 2011. In 2012, the linear relationship stopped at 112 DAP for NC; 107 DAP for TRBD and MLBD while for TR, ML, BD the linearity continued till maturity, that is 119 DAP (Figure 7.10). The discontinuity in linear relationship between TIPARs and DAB occurred when photosynthate translocation replaces radiation interception as the main factor for biomass production (Black and Ong, 2000). This departure of the linearity as the plant approaches maturity was also observed in Queensland (Sandana et al., 2012).

From 93 DAP in the first season and 98 DAP the second season, the crop began to show abrupt reduction in biomass production, a response to lower water availability during the reproductive stages in August and September, 2011 and October, 2012 respectively. However, reduction in biomass accumulation was delayed until 102 DAP in TRBD and MLBD. Reduction in biomass accumulation in 2012 started from 105 DAP in BD and NC but started earlier (98 DAP) in TR and other treatments. The delay was due to higher seasonal SWS in these treatments, which delayed stomata closure, lengthened seed filling and accumulation of biomass. The first season was drier with lower cloud cover but the seasonal average PAR was lower than that of 2012. Soil fertility and higher fIPARs contributed to accumulation of biomass in the first season. In addition, higher accumulation of biomass in 2012 in the treatments where water was conserved was because of the leaf area. Higher LAIs in the second season in the treatments where water was conserved led to greater interception of PARs during the reproductive stage of the crop and hence higher biomass accumulation. The cumulative instantaneous PARs ranged from 295 MJ m^{-2} for TR to 323 MJ m^{-2} for BD in the first season while in the second season, it ranged from 357 MJ m^{-2} to 446 MJ m^{-2}. Lower seasonal IPARs from SR is expected because average PARs for a day was used in the computation unlike seasonal instantaneous PARs where measurements were taken only during the peak periods of solar noon. RUEs for all the treatments in the two seasons by using PARs from the daily global SR and instantaneous measurements are in Table 7.7.

The RUE for TRBD was 1.11 g MJ^{-1} of IPAR in 2011 while BD had the peak RUE of 1.32 g MJ^{-1} in 2012. RUE for TRBD with seasonal SWS of 455 mm in 2011 was 5.4, 7.2, 24.3, 30.6 and 51.4% higher than RUEs for TRML, ML, BD, NC and TR respectively. However, in 2012, RUE for BD with seasonal SWS of 589 mm was 1.5, 6.1, 12.1, 16.7, 26.5 and 26.5% higher than RUEs of MLBD, ML, NC, TRML, TR and TRBD respectively.

Polynomial model fitted to RUE and dry biomass at harvest was not significantly correlated across seasons (BM (t ha^{-1}) = 58.5 - 126*RUE + 90.5*RUE^2 - 21.1*RUE^3), r^2 = 0.43, p = 0.12, SEE = 0.70 t ha^{-1}. This indicates that 43% of the variability in dry DAB at harvest can be explained by RUE. When the RUEs in Table 7.7 (column 3) and SWS in the two seasons were pooled together, there is no significant correlation between them (RUE = 2.26 - 0.001*SWS (mm)), r^2 = 0.06 and p = 0.40. This means that SWS account for only 6% of the variability in the RUE. Polynomial model fitted to the same data shows that there is no significant relation between RUEs and SWS under field conditions. For every increment of 20 mm in the SWS, RUE increased by only 0.02 g MJ^{-1}. This indicates that the rate at which Soybean uses SR to accumulate biomass depends to a lesser extent on the available soil moisture in the root zone. The average RUEs of the two approaches differed by only 0.65 and 0.63 g MJ^{-1} of IPAR in the first season and second seasons respectively (Table 7.7). The RUEs computed by the two approaches for all the treatments are significantly correlated (p = 0.01, r^2 = 0.48). This means that the two approaches gave appreciable similar RUEs for the treatments under field conditions in the two seasons.

When instantaneous PAR were used in computing RUEs, Peak seasonal RUE was 1.65 g MJ^{-1} of IPAR for TRBD that had the seasonal SWS of 455 mm in the first season. The RUE for TRBD was 14.5, 21.2 and 34.5% higher than those of ML, TRML and NC whose seasonal SWS were 445, 476 and 407 mm respectively. In the second season however, BD and ML whose seasonal SWS were 565 and 573 mm respectively had the peak RUE of 1.24 g MJ^{-1} of IPAR. Their RUEs was higher than those of TRML, MLBD, NC, TRBD and TR whose seasonal SWS were 548, 558, 562, 578 and 543 mm by 10.5, 12.9, 14.5, 22.6 and 24.2% respectively. Although the treatments where soil moisture was conserved had relatively higher RUEs compared with the convention practices NC, there was weak relationship between the SWS and RUEs (column 4, Table 7.7) when the data in the two seasons were pooled together. Seasonal SWS and RUEs were not significantly correlated at α = 0.05 (r^2 = 0.0005). Coefficient of variation CV of the RUEs in the two seasons was 0.18. A single cultivar of Soybeans was considered in this study and the seasonal average RUEs in the first and second seasons were 1.18 and 1.09 g MJ^{-1} of IPAR, respectively (Figure 7.9) with a reduction of about 7.63% in the second season.

There was no substantial difference in the seasonal RUEs of the crop in the two seasons despite higher TIPARs in the second season. Similarly, there was no significant difference in the means of the seasonal RUEs in the two seasons (p > 0.05) by using instantaneous PARs. These findings confirm the conservative nature of RUE for a particular environment reported (Monteith, 1994). In addition, the means of RUEs by using PARs from global SR in the two seasons are not significantly different (p > 0.05). The average RUEs for Soybeans in this study (Column 4) compare well with 1.46 and 1.99 g MJ^{-1} of PAR (de Souza et al., 2009). However, they are higher than the mean maximum RUEs of 1.20 g MJ^{-1} and two highest values of 1.26 and 1.15 g MJ^{-1} in six different studies in field experiment in Japan reported (Sinclair and Muchow, 1999). The RUEs of Soybeans is lower compared with other C_3 crops and this has been attributed to high energy contents protein and lipids, of the crop (Sinclair and Muchow, 1999).

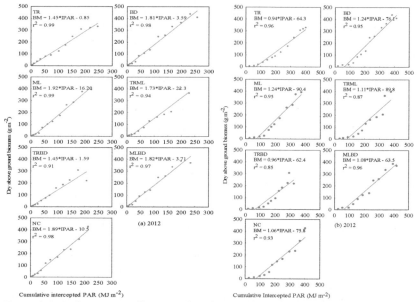

Figure 7.9. Radiation use efficiency of Soybean under rainfed conditions by using (a) cumulative PARs derived from Global SR and (b) PARs from instantaneous measurements (data after senescence of the plant at the late stage in each treatment were not included). Average of the dry above ground biomass for each day from four replicates was used to compute the RUE

7.7   Use of light model in simulating dry matter

The degree of agreement (d) between the measured and simulated biomass accumulation was excellent ranging between 0.98 for TR and 1.00 for TRBD in 2011 and 0.97 to 0.99 in 2012 (Figure 7.10). Measured and simulated biomasses were significantly correlated ($0.95 \le r^2 \le 0.97$; $24.8 \le SEE \le 43.8$ mm and $p < 0.05$) across the treatments and years. Similarly, in the second season, $0.75 \le r^2 \le 0.98$; $22.2 \le SEE \le 50.7$ mm. The assimilation model assumes a continuous and indefinite increment in biomass accumulation. In real plants this is not always true because after post-anthesis when the fruit had developed and fraction of the incident radiation intercepted by canopy had reduced considerably due to shedding of leaves, there will be a reduction in the radiation conversion efficiency. This may lead to reduction in the accumulation of biomass even at higher SR.

7.8   Discussion

The TRML that had the highest seasonal SWS of 476 mm in the first season had least λ of 0.40. Its λ reduced by 0.06 and 0.04 from those of NC and TR respectively. Similarly, it reduced by 0.08 from those of ML, TRBD and BD. In the second season, TRBD that conserved the highest SWS of 578 mm had λ of 0.50. It reduced by 0.05 for ML and 0.02 for ML. Pooled over the years, polynomial model fitted to SWS and extinction coefficient are not significantly correlated ($r^2 = 0.43$, $p > 0.05$, SEE = 0.03.).

This indicates that SWS accounts for 43% of the variability in the λ. The CV for the seasonal extinction coefficients is 0.08 and 0.05 for the first and second seasons respectively. Low CV and r^2 of λ in the growing seasons indicated that it was not affected substantially by water conservation practices, SWS and changes in weather conditions that occurred in the study area. Similarly, lack of significant difference in the λ in the two seasons showed that it was conservative for Soybeans.

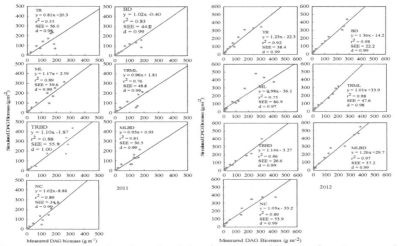

Figure 7.10. Measured and simulated dry above ground biomass by using a simple assimilation model under rainfed conditions

Data obtained in this study showed that the λ for the green leaf area (GLA) were higher in the treatments during early stage of growth than for developed canopy (Data not shown). This can be attributed changes in orientation of the leaves (canopy architecture and density during crop development), angles of inclination occurred mainly in response to the availability and intensity of SR, changes in solar zenith angles (Bonan, 2002). It was not affected maximally by water stress along the plant cycle especially in the first season. Differences in seasonal rainfall did not result in substantial changes in λ of the crop. Generally, a canopy with horizontal leaves have extinction coefficients ranging from 0.7 to 1.0 while those with more erect leaves have small values ranging from 0.3 to 0.6 (Atwell et al., 1999). Values greater than 1 indicates horizontal or regular leaf distribution. Since λ is a measure of light profile in a canopy and a function of LAI, higher LAIs in the treatments where water was conserved in the seasons resulted from lower λ. Seasonal λ are available for Soybeans in literature but none is reported for the crop under water conservation practice. Average seasonal λ in this study is relatively lower where combination of water conservation was used. Therefore, it can be presumed that the canopy is more erect under water conservation and causing the incoming radiation to be evenly distributed. This reduces the light intensity, which is intercepted by each plant, and ensures more even and uniform distribution and efficient light conversion.

Moisture availability in the root zone plays a prominent role in foliage development and growth of crops. Therefore, higher LAIs, which resulted into higher fIPARs in the treatments where water was conserved is attributed to higher SWS in the root zone. For instance, in 2011 when the seasonal SWS for TRBD was 455 mm, the peak fIPAR during seed filling stage was 0.91. It was higher than those of MLBD,

TRML and NC by 8.8, 12.1 and 13.2% respectively. In the second season, seasonal SWS of 548 mm for TRML resulted in the fIPAR of 0.95. It was higher than those of MLBD, TRBD and NC by 0.00, 8.42 and 7.37% respectively. Compared with NC, the fIPAR of the treatments where water was conserved were significantly higher.

Higher DABs in the treatments where water was conserved was due to higher SWS, which made water available for the crop and reduced greatly the loses due to runoff and evaporation. This is evident from the high correlation between SWS and DAB at harvest ($r^2 = 0.79$). Conservation of water in the stated treatments would result into higher HIs because moisture would be made available in the root zone of the plant during the sensitive seed filling stage of the crop where water stress could cause yield loss. The observed differences in the seasonal average dry biomass at harvest resulted from the variations in the environmental conditions, differential TIPARs and variations in LAIs. The fIPARs in 2011 in the growth stages were lower (Table 7.8) due to lower LAIs. Therefore, the reduction in accumulated biomass in the first season resulted from lower TIPARs not reduction in the RUE of the crop. This is clear from the good and higher correlation between DAB at harvest and IPAR ($r^2 = 0.76$) than dry DAB at harvest and RUE ($r^2 = 0.43$) in the pooled data across the seasons.

Reduction in the RUEs after R7 growth stage was due to the reduction in the photosynthetic capacity of the leaf during maturity of the crop and this observation was similar to other studies reported by (Sinclair and Muchow, 1999). Similarly, the high nitrogen required in the leaf makes it difficult for the crop to maintain a high photosynthetic rate even at high intercepted radiation (Sinclair and Horie, 1989) at late stage. However, decline in biomass accumulation in 2011, which led to reduced RUEs did not occur in all the treatments in 2012. This was due to optimal growth conditions from adequate rainfall in the second season compared to 2011.

The average seasonal RUEs (column 3, Table 7.7) for BD and TRBD where soil water was conserved differed from that of NC by 0.03 and 0.23 g MJ^{-1} of IPAR respectively. However, the RUEs for other treatments where soil water was conserved were lower than that of NC in the first season. In the second season however, the seasonal RUEs of only ML was higher than of NC by 0.03 g MJ^{-1} of IPARs while those of other treatment are lower. The results obtained in this study indicate that there were variabilities in the RUEs of a single cultivar of Soybean under water conservation practices although these differences are not significant. This finding further supports the recent suggestions that the assumption of a constant or highly conservative value of RUE within a specie or cultivar except when there is severe water stress may be incorrect (Bonhomme, 2000). The differences in the RUEs among the treatments were due to differences in their LAIs, IPAR and accumulated biomasses. In addition, the accumulated biomass also depends on available moisture in the root zone for transpiration.

The small differences in the seasonal average RUEs of Soybean in the two seasons could result from the differences in the observed meteorological conditions at the experimental fields. For instance, the average maximum and minimum seasonal temperatures in 2011 were higher than those of 2012 and this might be responsible for the differences observed in RUEs. The difference in seasonal maximum air temperatures in the two seasons was only 1.9°C (Appendix C) and not high enough to significantly affect RUEs of the crop when compared to the large variation observed in air temperature (19.9 to 39.5°C) in Ile-Ife during the cropping seasons. Although the air temperatures during the reproductive stage were higher in 2011 than 2012, this variation could result in the differences in the RUEs in the two seasons. VPD is a good indicator of the evaporative capacity of the air and a measure of the atmospheric drought by plants (Allen et al., 1998). It is a measure of the driving force for rate of transpiration in plants including Soybeans (Fletcher et al., 2007; Sadok and Sinclair, 2009a,b). The VPD was not very different in the two seasons, despite the differences in seasonal rainfall and air temperatures. The mean VPD during growing seasons in 2011 and 2012 were 0.64 and 0.62 KPa, respectively. The difference of 0.02 KPa was too low to significantly affect RUEs. These findings supports the argument of Arkebauer et al. (1994), that RUE cannot be expected to be constant, even within a single species or genotype, in the face of changes in other environmental variables.

It has been observed that diffuse radiation affects RUE of Soybean (Sinclair et al., 1992). Increase in fraction of diffuse radiation causes increase in accumulation of dry matter because of the contribution of the shaded leaves, which are more photosynthetically active than those exposed to direct sunlight and thereby contribute to increase in RUEs of the crop. Low TIPAR in 2011 explains why the average seasonal RUE was higher compared with RUE for the treatments in 2012. An increment of 6 to 33% in RUE has been attributed to the diffuse components of incident SR (Alton et al., 2007). During the period of cultivation of Soybean in Ile-Ife from May to October (rainy season), the degree of cloudiness was very high (visual observations) and this is a major characteristics of the weather conditions in Ogun-Osun River Basin, Nigeria. Cloudiness of the air makes a greater proportion of the incident radiation that reaches the ground surface as diffuse radiation, which probably contributed to the difference observed in the RUE in this study. Incidentally, lack of measured data on the field or record of diffuse radiation of Ile-Ife could not be used to substantiate the claim that the variability is responsible for the disparity in RUEs. Reduction in rainfall during the reproductive stage of the crop in the first season caused a reduction in dry matter and leaf area indices compared to the second season. This is a drought tolerance mechanism developed by the crop for conserving water. The closure of stomata in such a period of drought causes reduction in carbon dioxide fixation and consequently reduces the rate at which the IPAR was utilised in 2011. Water conservation practices, which retained more water in the root zone of the crop, can serve as a measure to maintain uninterrupted accumulation of biomass during short or prolonged period of drought in the study area.

7.9   Conclusion

Leaf area index, accumulation of biomass, extinction coefficient and radiation use efficiency of Soybean were measured for two seasons at the Teaching and Research Farms of Obafemi Awolowo University, Ile-Ife. There was weak correlation between extinction coefficients and seasonal SWS. Average extinction coefficient ranged from 0.40 for TRML to 0.55 for ML. Water conservation did not have significant effect on seasonal extinction coefficients of Soybean. The LAIs and fIPAR of the cultivar investigated reached the peak during seed filling stage of growth in the two seasons. This study show that water conservation led to significant increase in SWS and thereafter the LAIs among the treatments considered especially the Soil bunded (BD) plot where surface runoff was greatly minimised during the growing seasons and Mulch (ML) plot where evaporation was minimized by covering soil surface with plant materials. Seasonal IPAR and SWS are strongly correlated to DAB at harvest than RUE. Soil bunded and ML plots had higher cumulative DAB compared with the Control treatment (NC).This showed that the use of water conservation especially soil bund and mulches in the cultivation of the crop led to higher productivity of dry matter during the growing seasons in Ile-Ife. Accumulation of biomass is an important factor that determines yields in plants; the use of water conservation could be promising in increasing the seed yields and other yield components of Soybean especially when there are short or long fluctuations in rainfall.

By using PARs from the daily measurements of SR, seasonal average RUEs are 1.32, 1.69, 1.84, 1.87, 1.90 g MJ^{-1} of IPAR respectively for TR, TRBD, TRML, MLBD, and 1.92 g MJ^{-1} of IPAR for both BD and NC. With the use of instantaneous measurements of PARs with Ceptometer, average seasonal RUEs are 0.87, 1.07, 1.08, 1.18, 1.21, 1.27 and 1.31 g MJ^{-1} of IPAR for TR, NC, MLBD, BD, TRML, ML and TRBD respectively. RUE and seasonal extinction coefficient are conservative for Soybeans. There was very weak correlation between seasonal SWS due to water conservation practices and RUEs. Variations in the RUEs among the treatments were due to variabilities in accumulated biomass and IPARs when both instantaneous PARs and PARs from SR were used. Variation in RUEs was not due to environmental factors such as differences in VPD during the periods of this study. In situations where continuous or hourly measurements of either PARs or global SR are not available, instantaneous measurement of PARs near noon can be used to determine reliable data on RUEs of Soybeans in the study area. Extinction coefficient obtained for Soybeans in this study can be used in computing light interception and simulation of canopy cover of the crop in other areas within Ogun-Osun River Basin whose weather conditions are similar to that of the study area.

7.10   Soil water balance

7.10.1 Seasonal evaporation from the soil and transpiration from the plants

Figure 7.11 shows the trends in daily evaporation of water from soil and transpiration from crop emmergence till physiological maturity in 2011 and 2012 for the plants in the BD plot. Similar trends for other treatments are in Appendix H. At floweering stage in 2011 and 2012, transpiration constituted about 64% of the total evapotranspiration while during pod initiation and seed filling stages, transpiration increased to 74 and 89% respectively of the total evapotranspiration (Figure 7.11). This was due to the increament in CC (LAIs) as the crop developed after establishment to the mid-season in both years. Soil water evaporation reduced to the barest minimum of 19 and 8% respectively in 2011 and 2012 during seed filling stage and the peak transpiration was 92% of evapotranspiration at this stage of the growth. This is expected anyway because it is at this stage that solar energy captured from the sun was being assimilated into the seed and fruitful portion of the harvest. Shortly after seed filling, transpiration reduced to the barest minimum.

Table 7.8 shows the DABs at harvest, seasonal transpiration, evaporation, and evapotranspiration among the treatments in both seasons. Soil water evaporation was considerably higher for the treatments where water was conserved in the second season than in the first season. For instance, evaporation constituted 35.1% of the crop water use for control treatment in the first season but for TRBD, TRML and Soil bund BD, they were 50.2, 43.8 and 45.3% respectively. In the second season, comparative soil evaporation was 67.1, 70.8 and 69.3% for TRBD, TRML and BD respectively. Evaporation was higher in the treatments where water was conserved because of higher rainfall, which occasional resulted in surface ponding of water and subsequently evaporation to the atmosphere without contributing to the crop water use during the growing seasons. This is in agreement with the findings of (Cooper et al. 1983; French and Schultz, 1984; Siddique and Sedgley, 1986; Loss et al., 1997; Siddique et al., 1998) that cereals and grain legume crops can lose up to 60% of their crop water use to soil evaporation.

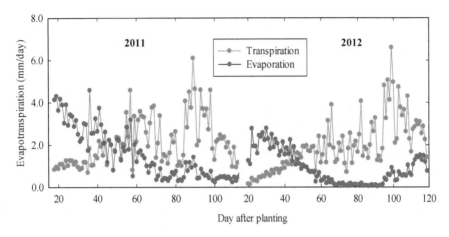

Figure 7.11. Daily soil evaporation and transpiration estimated by using the single crop coefficient approach and fIPAR radiation for the Soil bund (BD) under rainfed conditions

Table 7.8. Biomass at harvest, seasonal crop water use, water productivity for biomass production (WP$_{biomass}$) and transpiration efficiency (TE) in the rainy seasons

Year	Treatment	Biomass at harvest (t ha⁻¹)	STP$_r$ (mm)	SEP$_r$ (mm)	Crop water use (mm)	WP$_{biomass}$ (kg ha⁻¹ mm⁻¹)	TE (kg ha⁻¹mm⁻¹)
2011	TR	1.62	194	157	351	4.62	8.40
	BD	1.24	204	169	373	3.32	6.10
	ML	1.42	196	208	404	3.51	7.20
	TRML	1.25	190	148	338	3.70	6.60
	TRBD	2.11	202	204	406	5.20	10.4
	MLBD	1.02	202	109	311	3.28	5.00
	NC	1.01	200	108	308	3.28	5.10
	Average	1.38	198	158	356	3.84	7.00
2012	TR	2.96	173	389	562	5.27	17.1
	BD	3.25	181	408	589	5.52	18.0
	ML	3.11	195	338	533	5.83	15.9
	TRML	2.24	164	398	562	3.99	13.7
	TRBD	1.85	180	367	547	3.38	10.3
	MLBD	3.12	177	385	562	5.55	17.6
	NC	2.12	175	381	556	3.81	12.1
	Average	2.66	178	381	559	4.76	14.9

STP$_r$, SEP$_r$ are the seasonal transpiration and evaporation respectively under rainfed conditions. The DAB at harvest are the mean from four replicates of each treatment.

In the treatments where mulch was used to conserve moisture, there were considerable reductions in evaporation especially in the second season. In 2011, evaporation in MLBD reduced by 46.6% compared with TRBD. Similarly in 2012, evaporation in ML and NC reduced by 17.2 and 6.6% respectively compared with BD. Similar experiment conducted at Yucheng Stations with plastic film mulching showed that soil evaporation under canopy was reduced by 40 to 60% (Zhang et al., 1992). A light-decomposed plastic film used as mulch in cotton fields at the Nanpi experimental Station showed that soil evaporation was reduced by 50 to 72% (Wang et al., 1993). Lower water productivity in MLBD in the second season was due to higher proportion of the non-productive evaporation of water from soil. Seasonal transpiration for TRBD was 5.94 and 3.96% higher than those of the TRML and TR, respectively but 0.99% higher than that of NC in the first season. Similarly, seasonal transpiration in the second season for ML was 15.9 and 10.3% higher than those of TRML and NC, respectively. In the two seasons, the treatments where soil moisture was conserved had higher seasonal transpiration compared with the conventional practice (Table 7.8).

In the second season, despite higher evaporation of the soil moisture, biomasses at harvest were higher in the treatments where soil water was conserved compared with the similar treatments in 2011. Reductions in biomasses at harvest and $WP_{biomass}$ in NC and even few other treatments where water was conserved in 2011 was due to lower SWS and water stress which resulted to sudden closure of the plant stomata to conserve water as reported in (Passioura and Angus, 2010). $WP_{biomass}$ were higher in both seasons in the treatments where water conservation measures were applied except for TRBD in the second season. For instance, $WP_{biomass}$ for BD, ML, TRML, TR and TRBD were 1.2, 6.6, 11.4, 29.0 and 36.9% respectively higher than that of NC. Similarly, in the second season, $WP_{biomass}$ for TRML, TR, BD and ML were 4.5, 27.7, 31.0 and 34.6% respectively higher than for NC. In the two seasons, the average $WP_{biomass}$ for TRML, TRBD, MLBD, BD, ML and TR are 7.8, 17.4, 19.7, 19.8, 24.1 and 28.1% higher than that of NC. Linear model fitted to the SWU and DAB at harvest in the two season shows that they are significantly correlated (BM (kg ha^{-1}) = 0.006×SWU (mm) - 0.92; r^2 = 0.75; SEE = 0.43, p < 0.05). Polynomial model of the order 2 gave the same r^2. This means that for 75% of the variability in DAB can be explained by SWU. In addition, for every increment of 10 mm in the SWU of Soybeans, DAB at harvest increased by 0.06 kg ha^{-1} under field conditions. The least amount of water that evaporated from the soil during biomass production in the two seasons was about 133 mm (Figure 7.12a). Negative, and significant correlation was found between STP_r and TE with a linear model (TE (kg ha^{-1} mm^{-1}) = -0.25*STP_r (mm) + 58.9], r^2 = 0.47, p < 0.05, SEE = 3.63 kg ha^{-1} mm^{-1}) (Figure 7.12b). It indicates that for every increment 10 mm in STP_r, TE decreases by 2.5 kg ha^{-1} mm^{-1}.

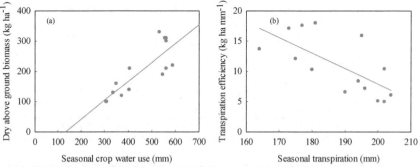

Figure 7.12. Production of (a) biomass in relation to a seasonal crop water use. The slope of the line represents average WP$_{biomass}$ for Soybean in Ile-Ife. The intercept on the x-axis denotes the minimum amount of water lost by evaporation from the soil surface. Reduction in (b) transpiration efficiency with seasonal transpiration under rainfed conditions

This result is similar to the findings for grain crops such as common Beans (White and Catillo (1990), Wheat (Siahpoosh and Dehghanian, 2012) and Zhang et al. (1998) who reported that water stress could reduce TE.

7.10.2 Seasonal transpiration efficiency (TE) and RUE

The pooled data of the TIPAR and TE in the two seasons are strongly and significantly correlated (TE (kg ha^{-1} mm^{-1}) = 0.13*IPAR (g MJ^{-1}) - 19.2), r^2 = 0.83, p < 0.05, SEE = 2.06). This indicates that IPAR accounts for 83% of the variability in TE for Soybeans under field conditions. For every increment of 10 MJ m^{-2} in the IPAR, there is an increase of 1.3 kg ha^{-1} mm^{-1} in the TE. Polynomial model fitted to the TE and IPAR gave a more strongly fitted relationship (r^2 = 0.90). These finding supports the facts that rate of transpiration for the production of plant matters is largely controlled by radiant energy (Mavi and Tupper, 2004). Polynomial model fitted to the pooled data of the RUE and TE shows that they are not significantly related (RUE = 3.90 - 0.60*TE + 0.05*TE^2 - 0.001*TE^3), r^2 = 0.20, p = 0.51, SEE = 0.25 g MJ^{-1} IPAR. This indicates that TE accounts for 20% of variability in RUE and that for every increment of 1kg ha^{-1} mm^{-1} in TE, there is a reduction of about 0.01 g MJ^{-1} in RUE.

Many factors are responsible for the variability in the seasonal WP$_{biomass}$ and TE. Available moisture in the root zone and proportion of diffuse radiation affect TE. Assimilation in diffuse radiation is greater than in an equivalent flux of direct radiation and that diffuseness of radiation has little effects on transpiration efficiency. The analysis carried out by Rodriguez and Sadras (2007) suggested that 1% increase in the proportion of diffuse radiation would lead to an increase of 0.5 kg ha^{-1} mm^{-1} in transpiration efficiency (Passioura and Angus, 2010). Rainy season in Ile-Ife is often characterised by diffusion of SR and this lead to higher biomass, WP$_{biomass}$ and TE. Similarly, rise in the atmospheric carbon dioxide level may lead to increase in transpiration efficiency of C3 crops such as Soybean. Incidentally, lack of data on diffuse radiation in the study area does not substantiate the fact that the observed differences in the WP$_{biomass}$ and TE of the crop was due to diffuseness of the incoming SR or variability in the carbon dioxide concentration in the growing seasons.

Effects of field management to reduce soil water evaporation were evident in the second season when minimum soil water evaporation was recorded for ML compared with other treatments in 2012. For instance, seasonal evaporation for TRBD and ML reduced from that of NC by 3.82 and 12.7% respectively. This supports the statement that the effects of field management on TE is very low compared with atmospheric conditions except that it reduces soil water evaporation (Passioura and Angus, 2010).

7.10.3 Soil water storage and dry above ground biomass at harvest

TRBD that had seasonal SWS of 455 mm produced peak biomass of 2.11 t ha^{-1} at harvest and was 52.1 and 23.1% higher than the biomasses of NC and TR respectively in 2011. In addition, the DAB at harvest for TRBD was higher than those of ML, TRML and MLBD by 32.7, 40.8 and 51.7% respectively. In the second season, the seasonal SWS for BD was 565 mm and its dry biomass at harvest was 31.1, 34.8% and 43.1% higher than those of TRML, NC and TRBD respectively. It was 4.3 and 4.0% higher than those of ML and MLBD, respectively. Pooled over the years, polynomial model fitted to SWS and DAB at harvest shows that they are significantly correlated (r^2 = 0.61, p < 0.05, SEE = 0.58 t ha^{-1}). This means that 61% of the variation in DAB at harvest can be explained by using SWS.

7.10.4 Water productivity for biomass (WP$_{biomass}$), Intercepted Photosynthetically Active Radiation (IPAR) and RUE

Higher seasonal average maximum and minimum relative humidity in 2011 is a reason for the higher WP$_{biomass}$ in 2012 among the treatments where water was conserved (Table 7.1). Average seasonal WP$_{biomass}$ was 6.19 kg ha^{-1} mm^{-1} when the biomasses in the two seasons were pooled together (Figure 7.12a).The difference in the seasonal average WP$_{biomass}$ for all treatments in the two seasons was 0.92 kg ha^{-1} mm^{-1}. The mean of the WP$_{biomass}$ for the two seasons are not significantly different, p = 0.08 at α = 0.05. These findings was similar to the results of Briggs and Shantz (1916) which shows that WP$_{biomass}$ is approximately constant for a single environment and increases with atmospheric humidity (Table 7.1). Management of soil to conserve water in the stated treatments resulted in improved (water productivity) water use efficiency for biomass production (WP$_{biomass}$) in the two seasons.

The polynomial model fitted to the pooled data of the WP$_{biomass}$ and TIPAR for the two seasons shows that they are significantly related (WP$_{biomass}$ (kg ha^{-1} mm^{-1}) = 51.9 - 0.49*IPAR + 0.002*IPAR^2 - 1.3*10^{-6}*IPAR^3), r^2 = 0.53, p < 0.05, SEE = 0.77 kg ha^{-1} mm^{-1}. It means IPAR accounts for 53% of the variability in the WP$_{biomass}$ and that an increment of 5 MJ m^{-2} in the IPAR will increase seasonal WP$_{biomass}$ by 0.10 kg ha^{-1} mm^{-1}. Pooled over the years, WP$_{biomass}$ and RUE are not significantly correlated (WP$_{biomass}$ = 100 - 195*RUE +129*RUE^2 - 27.9*RUE^3, r^2 = 0.17, p > 0.05, SEE = 1.02 kg ha^{-1} mm^{-1}). Only 17% of the variability in the WP$_{biomass}$ can be explained by RUE (Column 3, Table 7.7).

7.11   Soil water storage and crop evapotranspiration (SWU)

Figure 7.13 shows the SWS and SWU in the two growing seasons. Yearly changes and differences were evident for the SWS. In the first season, the SWS was 406 mm for TRBD while in the second season, it was 589 mm for BD. Differences in SWS were found among the six water conservation methods. In 2011, seasonal SWS in TRML, TRBD, and MLBD were 14.5, 10.5 and 6.7% higher than that of NC while those of TR, ML and BD were 10.0, 8.5 and 6.4% higher than that of NC. The second season was much wetter than the first season and coupled with higher proportion of the clay at the experimental field, the SWS were much higher in the treatments where water was conserved than for 2011. The effectiveness of conservation practices were also seen in the values of the SWS. For example, the SWS were 578, 573, and 565 mm for TRBD, ML and BD, respectively. Although, significant differences were not found in the SWS during seed filling stages in the first season in August, SWS was higher in TRBD and TRML than other treatments. In the second season, however, SWS for ML was 151±12.1 mm and significantly higher than those of other treatments in September during seed filling. Rainfall, nature of the soil and water conservation measures contributed to the higher SWS TRBD, and BD compared with NC in the second season. The experimental field used in the second season contained higher percentages of clay soil and retained more water as expected (Kirkham, 2004).

There were different distribution patterns in the SWU$_r$ in the two seasons. Both air temperature and seasonal rainfall contributed to the seasonal variability in the crop water use in the two seasons among the treatments as expected. Seasonal rainfall was 539 mm in the first season when seasonal average crop water use was 356 mm while in wetter second season, seasonal rainfall was 761 mm when seasonal average crop water use was 559 mm. For instance the first season, was slightly warmer with average seasonal air temperature of 32.9 °C while in 2012, the comparative air temperature was

30.9 °C. SWU_r ranged from 308 mm for direct sowing without water conservation practice to 406 mm for TRBD in the first season. In the second season, crop water use ranged from 533 mm for ML to 589 mm for BD. Seasonal SWS and crop water use in the two seasons are strongly and significantly correlated (SWU (mm) = -364 + 1.64*SWS (mm)), r^2 = 0.89, SEE = 37.6 mm, p < 0.05, as expected (Figure 7.13). Similarly, SWU and IPAR are significantly related as expected (SWU (mm) = -245 +2.93*IPAR (g MJ^{-1})), r^2 = 0.86, p < 0.05, SEE = 42.2 mm. Polynomial model fitted to the pooled data of the RUEs and SWU use shows that they are strongly and significantly correlated (SWU (mm) = -3815+5590*RUE - 1762*RUE^2), r^2 = 0.65, p < 0.05, SEE = 70.8 mm. The models accounts for 85 and 65% of the variability in the SWU_r in relation to the IPAR and RUEs respectively.

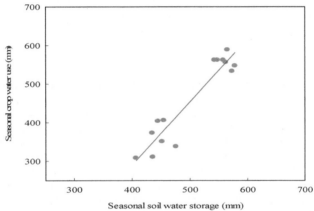

Figure 7.13. Seasonal soil water storages and crop water use under rainfed conditions

7.12   Soil water storage and seed yield

Figure 7.14 shows the seed yields and corresponding SWS for the six water conservation practices and the conventional method of cultivating the crop (control treatments). In the first season, TRML had the highest SWS of 476 mm; it produced a yield of 2.11 t ha^{-1}, which was lower than 2.95 t ha^{-1} for TRBD whose seasonal SWS was 455 mm. The seasonal SWS in TRML was 8.4, 8.61 and 5.04% higher than those of MLBD, BD and TR respectively. It was higher than the SWS of NC by 14.5%. Yield for TRML was lower than the yields for MLBD, BD and TR by 39.8, 17.5 and 20.4% respectively. However, the yield for TRML was higher than the NC and TR by 26.1 and 20.4%, respectively. Similarly, in the second season, TRBD had the highest water storage of 578 mm and was higher than TRML, MLBD and NC by 6.28, 4.18 and 3.35% respectively. However, the yield of 1.83 t ha^{-1} for TRBD was lower than those of MLBD, BD and ML by 77.6, 64.5 and 20.5% respectively. The yield was higher than the NC by 10.4%. Pooled over the years, seed yields and SWS in the two seasons were not significantly correlated by using linear model (Y (t ha^{-1}) = 1.58 + 0.001*SWS (mm)), p > 0.05, r^2 = 0.03). Seasonal SWS accounted for only 3% of the variability in seed yield. This indicates that increment in water storage is not the only determinant for seed yields in Soybean under tropical and sub-humid conditions in Ile-Ife. Other factors such as soil structure, soil fertility and weather conditions and differences in times of planting contributed to the variability in the seed yields observed in the two seasons.

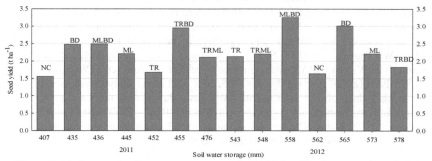

*Figure 7.14. Seasonal soil water storage and seed yields of Soybeans under rainfed
conditions. Each bar is the average of soil water storage from two replicates and seed
yields from four replicates*

7.12.1 Harvest index (HI), SWS and SWU

HI of 43.7±4.1% for the NC was the least of all the treatments in the first season but
was fairly uniform among other treatments where water conservation measures were
applied (Appendix J). The average seasonal HIs in the first and second season are 56
and 52% respectively. Pooled over the years, SWS and HI are significantly related (HI
(%) = -540 + 3.12*SWS - 5*10^{-2}*SMS^2+2.99*10^{-6}*SWS^3), r^2 = 0.55, p < 0.05, SEE =
3.23%. This indicates that 55% of the variability in the HI was due to SWS. HI and
SWU are weakly related (HI (%) = 84.7 - 0.35*SWU + 1*10^{-3}*SWU^2 - 9.12*10^{-7}*SWU^3), r^2 = 0.38, p > 0.05, SEE = 3.80%. Pooled over the years, HI and WP$_{seed}$ are
not significantly related (HI(%) = 343 - 20.7*WP$_{seed}$ + 0.42*WP$_{seed}$^{2} - 3*10^{-3}*WP$_{seed}$^{3}),
r^2 = 0.05, p > 0.05, SEE = 1.62%. This means that in the polynomial models, 38 and 5%
variabilities in the HI were due to SWU and WP$_{seed}$ respectively.

7.13   Yield, water productivity, harvest index and RUE

7.13.1 Seasonal crop water use and seed yield

Table 7.9 shows the yields, crop water use, WP, and HIs in the two seasons for the six
conservation measures and NC. Outputs of statistical analysis of the yields in both
seasons are in Appendices I, J and K. The seasonal crop water use (ET$_a$) ranged from
308 mm for NC to 406 mm for TRBD in the first season, while in the second season it
ranged from 534 mm for ML to 589 mm for BD. Crop water use in NC was lower
compared with the treatments where water was conserved. For instance, the ET$_a$ for
TRBD, TRML and MLBD were higher than that of NC by 24, 1and 0.9% in the first
season. However, those of TR, BD and ML were higher than that of NC by 12.3, 17.4
and 23.8 respectively. Compared with the first season, the ET in the second season
among the treatments were higher. However, the difference between them was lower
compared with the first season. For instance, ET$_a$ for TRML and MLBD were higher
than that of NC by 1.1% while that of TRBD was even lower than that of NC by 1.6%.
Higher seasonal rainfall of 761 mm and environmental conditions in the second season
contributed to higher ET$_a$.

The seed yields in the first season ranged between 1.56 t ha^{-1}for NC and 2.95 t
ha^{-1} for TRBD. In the second season, the comparative seed yields range between 1.64
for NC and 3.25 t ha^{-1}for MLBD, respectively. The average seed yield in the first season

was 2.21 t ha^{-1}, while in the second season it was 2.32 t ha^{-1}. Although, field was managed to conserve water in the cultivation of the crop and TRBD that had the peak SWS produced higher yields compared with other treatments in the first season, there was no significant difference between the seed yields for TRBD, TRML and MLBD at 5% level of comparison because it was higher by only 28% and 16% respectively. Similarly, the seed yields for BD and MLBD were not significantly different because the yield in MLBD was higher than that of BD by 0.4% in the first season and the difference in the seed yields between MLBD and BD in the second season was 7%. This indicates that the use of mulch with Side bund in the cultivation of the crop did not produce any significant benefit in terms of yields. The yield for TRML was not significantly higher than that of TR (20%) difference in the first season and 33% in the second season. This means that either treatment will produce similar yields in the cultivation of the crop. Significant differences in yields are found only between TRBD, MLBD and NC in the two years with the yields in TRBD and MLBD higher than that of NC by 45% and 49% respectively. The use of Mulch with Side bund was justified by significant increase in the yields compared with other treatments in the second season unlike in the first season when there was a yield gap of only 12%. A recent study show that the use of straw or organic mulch increased water productivity of tomato by 37% under rainfed farming (Mukherjee et al., 2010). The highest yield recorded for TRBD (Table 7.9) could be attributed to the conservation of water in the soil such that higher water storage was achieved compared with the NC (Table 7.4). Statistical analysis reveals that the average seed yields in the two seasons were not significantly different (p > 0.05) despite the fact that the second season was wetter. Pooled over the years, the polynomial model fitted to the seasonal crop water use and seed yield gave (Y (t ha^{-1}) = -34+0.26*SWU-0.0001*SWU^2)), r^2 = 0.26, p = 0.36 and SEE = 0.52 t ha^{-1}. This indicates that seasonal crop water use accounts for 26% of the variability in the seed yield. When considered across the years and water regimes, seed yield and TE were not significantly correlated by using polynomial model (Y (t ha^{-1}) = 0.08 +0.73TE - 0.08TE^2 + 0.003TE^3, r^2 = 0.46, p = 0.09, SEE = 0.44 t ha^{-1})).

7.13.2 Seasonal crop water use and water productivity for seed (WP$_{seed}$)

A lower seasonal crop water use in the NC resulted in its lowest WP$_{seed}$ of 5.06±1.85 kg ha^{-1}mm^{-1} (Table 7.9) compared with 7.26±0.73 kg ha^{-1} mm^{-1}and 7.99±2.85 kg ha^{-1} mm^{-1} for TRBD and TRML, respectively in the first season. Appendix J contains the output of statistical analysis of the WP in each year. Higher SWS also resulted in greater WP in 2012 compared with 2011. WP$_{seed}$ of 7.99 kg ha^{-1}mm^{-1} for MLBD was higher than those of TR, NC by 40.1 and 36.7% respectively, but 31.9 and 9.1% higher than ML, and TRBD respectively. This occurred because a greater proportion of the seasonal evapotranspiration was partitioned towards productive transpiration in the treatments where water was conserved in the first season (Table 7.9). WP$_{seed}$ was higher in the treatments where productive transpiration was higher than the non-productive evaporation. For instance, WP$_{seed}$ for TRBD was 7.26 kg ha^{-1} mm^{-1} when seasonal transpiration was 202 mm in the first season and 22.8% higher than WP$_{seed}$ for TRML when seasonal transpiration was 190 mm (Table 7.8). However, in the second season, peak seasonal transpiration of 195 mm resulted in WP$_{seed}$ of 4.15 kg ha^{-1} mm^{-1} and was lower than the peak water productivity for MLBD by 38.9%. Pooled over the years, the polynomial model fitted to the WP$_{seed}$ and seed yield indicates that WP$_{seed}$ account for 30% of the variability in the seed yields and that they are not significantly correlated (Y (t ha^{-1}) = 0.74 + 0.29*WP$_{seed}$ +0.03*WP$_{seed}^2$- 0.004*WP$_{seed}^3$), r^2 = 0.30, p = 0.29, SEE = 0.50 t ha^{-1}. Pooled over the years, polynomial model fitted to the WP$_{seed}$, STP$_r$ and SEP$_r$

shows that reduction in the WP_{seed} was due to reduction in seasonal transpiration rather than evaporation of water from the soil. This is evident in the higher correlation between WP_{seed} and STP ($WP_{seed} = -890 + 15*STP_r - 0.08*STP_r^2 + 1.57*10^{-4}*STP_r^3$, $r^2 = 0.54$, p < 0.05), than WP_{seed} and SEP ($WP_{seed} = 1.61 + 0.08*SEP_r - 4.00*10^{-4}*SEP_r^2 +5.47*10^{-7}*SEP_r^3$, $r^2 = 0.49$, p > 0.05).

Table 7.9. Yields and water productivity (seed) under rainfed conditions

Year	Treatment	Yield (t ha⁻¹)	Seasonal crop water use (mm)	Water productivity (kg ha⁻¹mm⁻¹)	Harvest index (%)
2011		Seed		Seed	
	TR	1.68 ± 0.50^{cb}	351	4.79 ± 1.42^{b}	47.4 ± 4.5^{bc}
	BD	2.48 ± 0.20^{ab}	373	6.65 ± 0.54^{ab}	51.5 ± 2.4^{ab}
	ML	2.20 ± 0.57^{abc}	404	5.44 ± 0.87^{ab}	53.2 ± 2.7^{ab}
	TRML	2.11 ± 0.72^{abc}	338	6.22 ± 2.12^{ab}	53.3 ± 4.1^{ab}
	TRBD	2.95 ± 0.30^{a}	406	7.26 ± 0.73^{ab}	57.6 ± 1.1^{a}
	MLBD	2.49 ± 0.89^{ab}	311	7.99 ± 2.85^{a}	56.7 ± 7.5^{a}
	NC	1.56 ± 1.56^{c}	308	5.06 ± 1.85^{b}	43.7 ± 4.1^{c}
	Average	2.21 ± 0.55	356	6.20 ± 1.68	51.9 ± 4.2
2012	TR	2.54 ± 0.82^{abc}	562	4.51 ± 1.46^{abc}	58.1 ± 2.3^{a}
	BD	3.01 ± 0.75^{ab}	589	5.11 ± 1.28^{ab}	53.1 ± 3.0^{a}
	ML	2.21 ± 0.44^{bc}	533	4.15 ± 0.82^{bc}	53.1 ± 3.9^{a}
	TRML	2.20 ± 0.73^{cb}	562	3.91 ± 1.29^{bc}	57.0 ± 1.1^{a}
	TRBD	1.83 ± 0.27^{c}	547	3.35 ± 0.49^{c}	56.4 ± 3.8^{a}
	MLBD	3.25 ± 0.52^{a}	562	5.76 ± 0.92^{a}	56.4 ± 1.4^{a}
	NC	1.64 ± 0.28^{c}	556	2.94 ± 0.51^{c}	57.9 ± 3.7^{a}
	Average	2.38 ± 0.58	559	4.25 ± 1.03	56.0 ± 5.1

Values are mean ± SD from four replicates. Means of the yield and WP_{seed} with the same letter are not significantly different at 5% (p > 0.05) level based on Duncan multiple comparison of means.

7.13.3 Water productivity, seed yield and biomass at harvest

Pooled over the years, WP_{seed} and seed yields are not significantly correlated (Y (t ha⁻¹) $= 0.74 + 0.29*WP_{seed} +0.03*WP_{seed}^2 - 0.004*WP_{seed}^3$, $r^2 = 0.30$, p > 0.05, SEE = 0.50 kg ha⁻¹ mm⁻¹). It means that 30% of the variability in the seed yield is accounted for by WP_{seed}. This implies that WP_{seed} cannot be increased without compromising seed yield under field conditions for Soybeans. Increment in WP_{seed} beyond a threshold of 8.1 kg ha⁻¹ mm⁻¹, seed yield begins to decrease. However, $WP_{biomasss}$ and biomass at harvest are strongly and significantly correlated by using both linear and polynomial models (B(t ha⁻¹) $= -25 + 17.7*WP_{biomass} -3.93*WP_{biomass}^2 + 0.30*WP_{biomass}^3$, $r^2 = 0.81$, p < 0.05, SEE = 0.40 t ha⁻¹). 81% of the variability in DAB at harvest is attributed to $WP_{biomass}$. This implies that it is possible to increase $WP_{biomass}$ without compromising biomass formation. Pooled over the years, WP_{seed} and $WP_{biomass}$ are not significantly related by using a polynomial model ($WP_{seed} = 169 - 112*WP_{biomass} + 24.8*WP_{biomass}^2 - 1.79*WP_{biomass}^3$), $r^2 = 0.30$, p > 0.05, SEE = 1.39 kg ha⁻¹ mm⁻¹. This means that 30% variability in the WP_{seed} is traceable to $WP_{biomass}$. In addition, this indicates that efficient use of water for formation of total above ground materials does not result into efficient use of water for seed formation under field conditions in the crop.

7.13.4 Seed yield, IPAR and RUE

Polynomial model fitted to the pooled data of the TIPARs and seed yields in the two seasons shows that they are significantly related (Y (t ha^{-1}) = -187+2.47*IPAR - 0.011*IPAR^2 + 0.001×10^{-2}*IPAR^3), r^2 = 0.61, p = 0.02, SEE = 0.38 t ha^{-1}). This means that the model accounts for 60% of the seed yield with respect to IPARs and that for every increment of 1 MJ m^{-2} in the IPAR, the seed yield increases by 0.01 t ha^{-1}. However, the pooled across the seasons RUEs (Column 3, Table 7.7) and seed yields are not significantly correlated (Y (t ha^{-1}) = 0.57*RUE (g MJ^{-1} IPAR + 1.29); r^2 = 0.07, p = 0.36. This means that the RUE accounts for only 7% of the variability in the seed yield at harvest. The linear model suggests that for every increment of 0.5 g MJ^{-1} of RUE, the seed yield will increase by 0.29 t ha^{-1}. Polynomial model fitted to the same seed yields and RUEs shows that they are not significantly correlated. These finding further disproves the argument that few evidence exists that incident radiation is a critical limiting factor determining crop growth under normal field conditions. Demetriades-Shah et al. (1992) advocated that analysis of crop growth in terms of cumulative intercepted radiation and the conversion efficiency of solar energy during dry matter production should be approached with caution because crop growth and productivity depends on soil, atmospheric, and biological factors, of which radiation is only one component. High and significant correlation between the seed yield and TIPAR for Soybeans further supports the validity, robustness and generality of the correlation between IPAR, growth and the conservative nature of RUE (Monteith 1994). The average seed yields (land productivity) of 1.60 t ha^{-1} for NC and 2.87 t ha^{-1} for MLBD in both seasons compare well with 2.7 t ha^{-1} (Bhatia et al., 2008) and 1.7 to 2.3 t ha^{-1} at the experimental sites of IITA in Nigeria in 1992, Benin, Ghana and Togo (Tefera, 2011). It was higher than 2.49 t ha^{-1} obtained at the experimental stations of the Institute of Agricultural Research and Training (IART) Ibadan, Nigeria (Adeniyan and Ayoola, 2006). It is far above the 1.5 t ha^{-1} recorded for the same cultivar in, Ilorin-southern Guinea savannah of Nigeria (Akande et al., 2007).

7.13.5 Water productivity for seed, IPAR and RUE

In the first season, WP$_{seed}$ for MLBD, TRBD, BD, TRML and ML, were 36.7, 30.3, 23.9, 18.6 and 6.99% higher than the water productivity for NC. WP for MLBD was significantly higher (p < 0.05) than those for TR and NC. WP$_{seed}$ of Soybeans for MLBD, BD, TR, ML, TRML and TRBD were 49.0, 42.5, 34.8, 24.8 and 12.2% respectively higher than of NC where no water conservation was used in the second season. Similar to the first season, WP$_{seed}$ for MLBD in the second season is significantly higher (p < 0.05) than that of NC. Polynomial model fitted to the pooled data of WP$_{seed}$ and TIPAR in the two seasons shows that the two are strongly correlated (WP$_{seed}$ = -564 + 7.44*IPAR - 0.03*IPAR^2 + 0.00*IPAR^3), r^2 = 0.71, p = 0.01, SEE = 0.89 kg ha^{-1} mm^{-1}. The model accounts for 71% of the variability in WP$_{seed}$ with respect to TIPAR and that every increment of 10 MJ m^{-2} of IPAR will increase WP$_{seed}$ formation by 0.10 kg ha^{-1}mm^{-1}. The polynomial model fitted to the pooled data of the WP$_{seed}$ and RUEs in the two seasons shows that they are not significantly correlated (WP$_{seed}$ (kg ha^{-1} mm^{-1}) = 30.1 - 35.1*RUE + 11.7*RUE^2), r^2 = 0.34; p = 0.1, SEE = 1.29 kg ha^{-1} mm^{-1}. This means that the model accounts for 34% of the variability in the WP$_{seed}$ due to RUE. WP$_{seed}$ formation in this study fall within the range of 1.37 to 7.88 kg ha^{-1} mm^{-1} for Soybean under rainfed conditions (Das, 2003) and higher than mean productivity of 1.27 to 1.28 kg ha^{-1}mm^{-1} in Obalum (2011) for mulch and unmulched treatments of Soybeans in Nigeria.

7.13.6 Harvest index, IPAR and RUE

HI of 43.7±4.1% for the NC was the least of all the treatments in the first season but was fairly uniform among other treatments where water conservation measures were applied. In the second season, the HIs were fairly uniform among all the treatments including NC (Table 7.9). The average seasonal HIs in the first and second season are 52 and 56% respectively. Pooled over the years, the HIs and TIPARs shows that they are not significantly correlated (HI = -70.9 + 0.99*IPAR - 0.00*IPAR^2), r^2 = 0.35, p = 0.09, SEE = 3.70%. Similarly, polynomial model fitted to the pooled data of the HIs and RUEs shows that they are not significantly correlated (HI (%) = -211 + 427 - 217*RUE^2 + 34.5*RUE^3), r^2 = 0.38, p = 0.17, SEE = 3.79%. These mean that 38 and 35% of the variabilities in HIs are due to RUE and IPAR respectively. In addition, when considered across years and water use, seed yield and HI were not significantly correlated (Y (t ha^{-1}) = 168 -10.3HI + 0.21HI^2 - 0.001HI^3, r^2 = 0.30, p = 0.30, SEE = 0.50 t ha^{-1}).

7.14   Partitioning of the dry above ground biomass at harvest

In 2011, TRBD had the highest seed yield of 2.95 t ha^{-1}, the proportions of the chaff and stem were 28.0 and 16.1% respectively while the seed constituted 55.9% (Figure 7.15). For NC, which had the lowest yield of 1.56 t ha^{-1}, the proportions of the chaff and stem were 31.7 and 27.8% respectively while that of the seed was 40.5%. The proportions of stems ranged from 16.1% for TRBD to 27.8% for NC, while for chaff it ranged from 24.8% for MLBD to 33.1% for TR. However, the proportion of seed ranged from 40.5% for NC to 55.9% for TRBD. The average proportions were 20.0, 30.3 and 49.7% for dry stem, chaff and seed respectively for all the treatments. However, in 2012 when the seasonal rainfall was higher, MLBD had the highest seed yield of 3.25 t ha^{-1}. The proportions of the dry stem and chaff were 29.3 and 19.9% respectively, while that of the seed was 50.9%. NC had the minimum seed yield of 1.64 t ha^{-1}, the proportions of stem, chaff and seed were 28.7, 24.7 and 46.6% respectively. Seasonal average of stem, chaff and seed were 31.2, 21.8 and 46.9% respectively for all the treatment. Higher proportion of the stem in the second season could be due to the higher precipitation compared to the previous season.

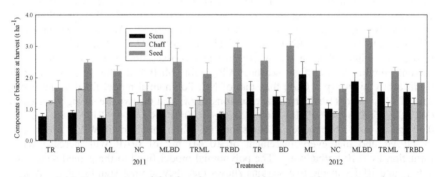

Figure 7.15. Partitioning of the dry above ground biomass at harvest into dry stems, chaffs, and seeds after harvest in the rainy seasons. Each bar represents the mean with standard error from four replicates

When the proportions of the seeds, chaffs and stems in all the treatments in the two seasons were pooled together, linear regression shows that there was no significant correlation between stem and chaff (r^2 = 0.13; SEE = 0.02; p = 0.21), seed and chaff (r^2 = 0.09, SEE = 0.52; p = 0.30) and seeds and stem (r^2 = 0.08, SEE = 0.52; p = 0.32). In the two seasons, there was weak but positive correlation between proportions of dry chaff and seeds as expected. This means that increase in the proportion of the chaff did not produce corresponding increase in seed yields and vice versa. Similarly, increase in stems did not produce proportionate increase in chaffs of Soybean.

7.15 Economic evaluation

7.15.1 Land limiting conditions

Higher TE and WP do not make any meaning if it does not translate into higher revenues and economic benefits in agricultural activities. Average seasonal costs of production per hectare for TR, MLBD, TRML and TRBD were 28.9% higher than those for ML and NC but about 10.1% higher than for BD. Increased cost of production for the treatments where water was conserved especially TR, MLBD, TRML and TRBD was due to the cost of making the Soil bunds and ridges and their regular repair after rainfall. Average revenue for MLBD was 4.3, 16.6 23.3, 24.5 26.4 and 44.2% higher than those for BD, TRBD, ML, TRML, TR and NC respectively. Incidentally, none of the treatments offered financial benefits after the cultivation. This is attributed to higher cost of production especially where water was conserved and low market price as at the time of harvest. The maximum gross revenue was 1.63×10^3 US$ for MLBD (Figure 7.16). Thus, the use of MLBD seems to be appropriate when land is limiting. However, ML had the minimum economic loss, which corresponds to the average seasonal crop water use of 469 mm. See Appendix L for the items.

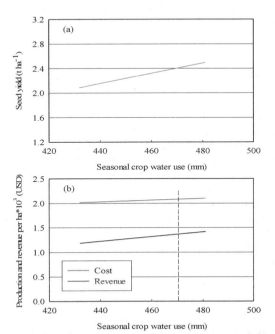

Figure 7.16. (a) Seed yield versus seasonal crop water use and (b) production cost versus seasonal water use in land limiting conditions

7.15.2 Water limiting conditions

Under water limiting conditions, any practice that generates the highest income per unit of water used is considered the most productive (Ali, 2010). The $WP_{economic}$ was higher in the treatments where water was conserved (Table 7.10). For instance, $WP_{economic}$ for TR, ML, TRML, TRBD, BD and MLBD were higher than that of NC by 14.1, 16.6, 21.0, 24.6, 32.0 and 41.8% respectively. $WP_{economic}$ of 3.01 US$ ha^{-1} mm^{-1} (0.30 US$ m^{-3}) for TRBD was higher than those of TRML and TR by 4.52 and 12.3% respectively. For MLBD, $WP_{economic}$ of 3.90 US$ ha^{-1} mm^{-1} (0.39 US$ m^{-3}) is higher than those of BD and ML by 14.5 and 30.3% respectively (Table 7.10). MLBD seems to be more appropriate under water limiting conditions. Greater land, water and economic productivity in the treatments where water was conserved justified the advantages of the methods over the conventional practices.

Table 7.10. Economic analysis of the production of Soybean under rainfed conditions

Treatment (1)	Price per tons (US$) (2)	Average yield (t ha^{-1}) (3)	Production cost ha^{-1} (US$) × 10^3 (4)	Revenue ha^{-1} (US$) (2×3) × 10^3 (5)	Benefit/Loss (US$) × 10^3 (4-5) (6)	Economic productivity (US$ ha^{-1} mm^{-1}) (7)
TR	567	2.11	2.28	1.20	-1.08	2.64
BD	567	2.75	2.05	1.56	-0.49	3.33
ML	567	2.21	1.62	1.25	-0.37	2.72
TRML	567	2.16	2.28	1.23	-1.05	2.87
TRBD	567	2.39	2.28	1.36	-0.92	3.01
MLBD	567	2.87	2.28	1.63	-0.65	3.90
NC	567	1.60	1.62	0.91	-0.71	2.27

7.16   Discussion

Soil moisture and temperature

The use of mulch increased the soil water content in the upper 30 cm of the soil by 17% while the use of Tied ridge increased it in the lower 30 to 50 cm by 22% compared with NC. Increase in the soil water in the treatments where water was conserved was due to the presence of plant materials as protective layers and the manipulation of the soil to reduce or eliminate runoff. Since a greater proportion of the root of Soybeans is located in the upper 30 cm of the soil, the use of water conservation practices will reduce moisture stress that could reduce the performance of the crop in case there is a drought due to sudden reduction or fluctuation of rainfall in the study area.

Higher temperature in the upper 30 cm of the soil at the early stage of growth of the crop in the NC was due to the exposure of the soil to the direct SR. The use of mulch reduced soil temperature in the upper 30 cm between 2.2 to 2.9°C compared with NC, TR, and TRML.

SWS, SWU and WP

Variability in the SWS, yield, crop water use and WP of the crop during the growing season can be attributed to two factors. These are water conservation practices and environmental conditions in the study area. Rainfall during seed filling in 2011 and

2012 constituted 32 and 45% respectively of the seasonal effective rainfall. Significantly low rainfall during this period in 2011 led to early leaf senescence and shortness in the period of seed filling inspite of the fewness of days and consequently reduction in the yields. This is similar to the finding of de Souza (1997) and Brevedan and Egli (2003) that a relatively short period of stress can lead to a disproportionate reduction in yield of Soybean. Higher maximum air temperature of 39.5 °C in 2011 compared with 31.1 °C in 2012 resulted in reduction of the duration of the seed filling stage (R5 to R8) compared with 2012 (Table 6.1). Although the difference in the duration of seed filling in the two years was three days, it could lead to reduction in the transfer of assimilate to the seed, which would eventually lead to the reduction in the yields in the treatments compared with those measured in 2012 (Egli, 2004).

The construction of side bunds resulted in trapping of higher proportions of rainfall in the two seasons compared with NC and hence led to increased SWS. Trapping of water within the root zone benefited the crop and the use of Side bund plus either Mulch or Tied ridge was efficient in conserving soil water and resulted in higher HIs in the treatments. Effects of water conservation practices were evident in increasing the SWS and STP_r in the current study. For instance in 2011, the SWS of 455 mm for TRBD resulted in STP_r of 202 mm (Table 7.8) and peak seed yield of 2.95 ± 0.30 t ha^{-1} (Table 7.9). Similarly, in 2012, SWS of 573 mm for ML produced STP_r of 195 mm and seed yield of 2.21 t ha^{-1}, which was not significantly higher than the yields in BD, TRML, and TR (Table 7.8).

Seed yield, IPAR, IPAR and HI

Seed yield of a crop can be expressed (Saha et al., 2012) as:

$$Y = IPAR \times fIPAR \times RUE \times HI \tag{86}$$

where IPAR, fIPAR, RUE and HI are as defined previously. Increasing any of these variables will increase the seed yield. Assuming RUE is conservative as obtained in the current study and HI does not change substantially, then the seed yield will depend largely on IPAR and fIPAR, which are functions of environmental conditions. Any process or condition imposed on the crop to increase its LAI will also increase fIPAR. The TIPAR will also increase due to increment in the foliage size and therefore increase the seed yield. This is true as it was found in the current study that there is higher and significant correlation between seed yield and TIPAR ($r^2 = 0.61$) than RUE ($r^2 = 0.07$) and HI ($r^2 = 0.30$) under field and rainfed conditions. Based on this, the crop should be cultivated in the study area when both rainfall and SR are optimum in order to ensure maximum land productivity.

In water-limited environment, a useful framework for evaluating the effects of grain yields is to express grain yield as the product of three largely independent entities (Westgate et al., 2004; Passioura and Angus, 2010) and expressed in the equation:

$$Y = T \times TE \times HI \tag{87}$$

where:

Y = grain yield (t ha^{-1})
T = seasonal transpiration (mm)
TE = transpiration efficiency (kg ha^{-1} mm^{-1})
HI = harvest index (%)

Increasing any of the three parameters will increase the yields of Soybean when water is limiting at any stage of growth of the crop (Westgate et al. 2004). For instance, since yield is directly proportional to transpiration provided TE and HI remains fairly constant, management practices such as mulching and decrease in tillage practices that increase the amount of soil water storage that can be utilized during later stages of crop development will increase the amount of water available for transpiration. Therefore, increase in seed yields in the treatments where water was conserved was due to higher correlation between seed yield and TE (r^2 = 0.46) rather than HI (r^2 = 0.30) and STP$_r$ (r^2 = 0.15). Similarly, increase in seed yields in the treatments where water was conserved was due to higher seasonal crop water use and WP$_{seed}$ rather than seasonal transpiration. This is evident in the better although low correlation between the pooled data of the WP$_{seed}$ and seed yield (r^2 = 0.30); seasonal crop water use and seed yield (r^2 = 0.26); than seasonal transpiration with seed yield (r^2 = 0.20). Reduction in the WP$_{seed}$ without much change in the seed yield was due to higher correlation in the pooled data between seasonal crop water use and WP$_{seed}$ (r^2 = 0.49) rather than WP$_{seed}$ and seed yield (r^2 = 0.30). The variation in TE could be interpreted based on photosynthetic capacity and stomatal conductance of the plants under varying SWS (Martin et al., 1994). Theoretically, more accessible water is conducted to more photosynthetic capacity, and thus, more gas is exchanged in the plants. Increasing conductance from the stomata and boundary layer may have enhanced transpiration per unit leaf area and reduced the TE. However, the smaller LAI in the treatments was a consequence of a growth reduction mechanism of tolerance, which reduced light interception and may increase the TE.

Variations in the environmental conditions in the two seasons also contributed to differences in the average seasonal yields. Although, all the treatments were subjected to the same environmental conditions in each season, amount of water stored in the root zone and transpired by canopy differed. Higher yields in the treatments where water was conserved were a result of increased meteorological components of TE. Water vapour concentration inside the leaf is saturated at any given temperature, and increases as temperature increases. Increase in temperatures results in higher vapour pressure deficit. Since TE is inversely proportional to VPD, therefore, TE will be higher under relatively cool and humid conditions (Westgate et al., 2004). Higher rainfall coupled with higher relative humidity in the rainy season of 2012 compared with 2011 resulted in elongation of the duration of seed filling and hence higher TE (Table 7.8). Despite the environmental conditions, the average seasonal seed yield only increased by 7%. Reduction in the seed yields even in the treatment where water was conserved especially in the first season was due to reduction in the IPAR than the RUE and SWS. This is evident from the good correlation between seed yield and IPAR (r^2 = 0.61) than seed yields and RUE (r^2 = 0.07), seed yield and SWS (r^2 = 0.02) and seed yield with SWU (r^2 = 0.26) in the pooled data over the years.

Increasing HI in Equation 85 is another way of increasing yields of Soybeans under rainfed conditions. The period and duration of the drought determine the effects it will have on HI. Low increase in the seed yields over the years among the treatments without much increase in the HI was due to higher correlation between seasonal crop water use and HI (r^2 = 0.38) rather than WP$_{seed}$ and HI (r^2 = 0.30). Low correlation between seasonal water use and seed yields over the seasons in this study clearly indicates that relative decrease in the seed yields is lower than relative decrease in the seasonal crop water use. This means that water shortage at sensitive stages of growth such as flowering and seed filling will have effects on the productivity of the crop.

Practices such as high population density of plants which increases canopy cover and reduces soil water evaporation, minimum tillage, disruption of hard pans, residue management and water conservation techniques that increase soil water infiltration such as Mulch, Soil bund, Tied ridge in this study will prevent premature senescence that are associated with drought (Westgate, 2004). The water conserved during vegetative stage could be used during seed filling and thereby resulting in increased HI. Higher rainfall and high humidity in 2012 benefited the crop and the use of Side bund plus either Mulch or Tied ridge was efficient in conserving soil water and resulted into higher HIs in the treatments. Longer duration of the seed filling in 2012 coupled with lower air temperature resulted in higher partitioning of assimilates to seeds than the vegetative part and eventually led to the higher yields compared with 2011. The differences in the yields among the treatments in the same seasons were also due to variability in the soil nutrients. This supports the findings that water in the soil-plant-atmosphere continuum water is not the only factor responsible for growth, development and high yields in Soybean (Egli, 2004).

Selecting the best conservation practice (s) among the six used in the current study is important. The factors considered in assessing the performances of the conservation practices are the average seasonal crop water use (Table 7.8), TE (Table 7.8), seed yields (Table 7.9), green WP (Table 7.9) and cost of production (Table 7.10). All the six conservation practices had higher average seasonal evapotranspiration than the NC. The average seasonal TE of the crop for the conservation practice with the minimum water use was 23.9% higher than that of NC. The minimum average seed yield of 2.11 t ha^{-1} for TR and maximum seed yield of 2.87 t ha^{-1} for MLBD were 24.2 and 44.3% respectively higher than for NC. The average minimum green WP$_{seed}$ of 4.65 kg ha^{-1} mm^{-1} for TR was higher than that of NC by 14% while 6.88 kg ha^{-1} mm^{-1} for MLBD was 41.9% higher than for NC. Aside ML that had the same cost of production with NC, the cost of production for other conservation measures were 28.9% higher than that of NC and this greatly increased the financial loss after sale to 24.5% for BD and 65.7% for TR. WP$_{seed}$ is more correlated to seasonal IPAR than both seasonal RUEs and SWUs.

In water limited environment, management practices that ensure highest land and WP under both rainfed and irrigation conditions are always advocated. The treatments where water was conserved had higher SWS, TE, seed yields and WP (biomass and seed) compared with the NC. MLBD had the maximum seed yield, WP$_{seed}$ and revenue. In addition, the use of Soil bund is also promising because the yields and revenue for MLBD were 4.29 and 4.18% higher. Although the construction of soil bund is laborious especially when done manually by the peasant farmers, the yields and WP at the end of the growing period justifies the effort made. The use of any of the six conservation measures will be useful in conserving soil water and reduce yields gaps in Soybean in Ile-Ife. Soybean growers in the study area are more interested in the practice(s) that will enhance their income and retain water in the soil for subsequent seasons, and hence ensure the use of land resources on sustainable basis. This innovative water conservation measure can be a promising means for resilient and adaptive cultivation of Soybean and thus reduce the yield gaps and ensuring sustainability of land and water resources under fluctuating rainfall and weather conditions in Ile-Ife and other agrarian communities in Ogun-Osun River Basin. Water conservation practice could be used to conserve soil water and increase water productivity of Soybean.

7.17 Conclusion

The evapotranspiration components, yield, water productivity for biomass and seed productions and HIs of Soybeans for six water conservation techniques and conventional cultivation method were studied for two consecutive rainy seasons. SWS depended on the nature of the soil and was higher where water was conserved compared with the conventional practice. It partitioned higher proportion of the crop water use towards transpiration and eventually increased the DAB, TE, seed yields and WP of Soybean under rainfed conditions in Ile-Ife. In Ile-Ife, minimum evaporation from the soil under rainfed conditions in the cultivation of Soybean was 133 mm. Soil water evaporation could be reduced below 133 mm by covering the entire soil under the plant. This practice although laborious but feasible in the study area because of abundant plant materials, which can serve as mulch.

DAB at harvest is well correlated to both seasonal soil water storage and crop water use. Similarly, seasonal water storage and crop water use are significantly correlated. Seasonal soil water storage and crop water use are not significantly related to seed yield of Soybeans. This is an indication that the crop is drought tolerant and application of water to the crop at sensitive stages of growth will lead to better use of water in cultivating the crop. Seasonal water use and water productivity are not significantly correlated. Harvest index is more correlated to seasonal water storage than seasonal water use. Water productivity is more correlated to seasonal transpiration than evaporation of water from the soil, which did not contribute to yield formation.

Transpiration efficiency is strongly sensitive to TIPAR and RUE. Seasonal water use of the crop depends on environmental conditions such as temperature and humidity and is significantly related to seasonal IPAR and RUE. Seed yield, water productivity of the crop increased with water conservation especially with Tied ridge plus Soil bund and Mulch plus soil bund. This was due to protective cover provided by the Tied ridge and Soil bund, which eliminated surface runoff and concentrated the available water in the root zone of the plants.

Seed yield is not significantly related to RUE of the crop. Water productivity for seed yield and biomass are significantly related to TIPAR but not significantly related to RUE. Harvest index of the crop increased considerable under humid conditions and water conservation practices. However, HI of the crop is not significantly related to IPAR, RUE, seasonal crop water use and water productivity for seed formation.

Erosion and loss of soil nutrients, which arise from incidental torrential rainfall, could be greatly reduced by Tied ridge and creation of Soil bund around plots on the field. The use of water conservation could be used to conserve soil water and reduce yield gaps in Soybean in sub-humid tropical weather of Ile-Ife. Although the cost of production of Soybean was higher than that of the conventional method, increased land and water productivity justified the effort made in conserving water. These innovative water conservation measures could be promising means for resilient and adaptive cultivation of Soybean in the study area and other parts of the world with similar climatic conditions. Through these means, sustainability of land and water resources under fluctuating rainfall and weather conditions in Ile-Ife and other agrarian communities in the Basin could be ensured. Mulch plus Soil bund or Soil bund are hereby recommended for the production of the crop.

8 Effects of deficit irrigation on soil water balance, yield and water productivity of Soybeans

The need for reduction in water use by agriculture is being advocated globally due to stiffer competition among fresh water users such as irrigation, industry, domestic and the environment. Several suggestions have been made by stakeholders in the irrigation sector to optimize the use of water for crop production. One of the suggestions is that water should be applied to crops when they need it most or when shortage of water could lead to significant reduction in yield. This approach is called regulated, pre-planned deficit evapotranspiration or DI (English et al., 1990). Articles have been published on the possibility of saving irrigation water through DI without significant reduction in the yield. For instance, Cowpea has the ability to maintain seed yield when subjected to drought during the vegetative stage provided that subsequent irrigation will not exceed eight days (Ziska and Hall, 1983). Several researches have been carried out on potatoes (Minhas and Bansal, 1991) to ascertain the possibility of achieving optimum yield under DI by allowing a specific level of yield loss of a crop by diverting irrigation water to another crop that has higher economic returns. Researches on identifying the critical stage where water stress can reduce yield and performance of Soybean are in progress. Available data show that an equivalent or greater yield can be obtained by delaying irrigation until Soybeans are in the reproductive stage of growth compared with the seasonal irrigation scheduling (Brady et al., 1974; Ashley and Ethridge, 1978; Martin et al., 1979; Heatherly, 1983; Elmore et al., 1988; Klocke et al., 1989; Specht et al., 1989). Stegman et al. (1990) stated that a short period of water stress during flowering may lead to a drop in flowers and pods at the lower canopy, but this will be compensated for by increased pod set at the upper nodes when irrigation resumes later in the crop life. Stegman et al. (1990) concluded that water stress in the full pods to the seed fill stage was most detrimental to yield in Soybeans. Ashley and Ethridge (1978) reported that irrigation during the vegetative growth only enhances canopy growth without corresponding increase in seed yield. Soil water stress at the vegetative stage reduces canopy growth and expansion, which in turn reduces both the dry matter yield and seed yield in Soybean (Constable and Hern, 1978; Sivakuma and Shaw, 1978). It was reported by Constable and Hern (1978) that water stress at the seed filling stage induced leaf senescence and greatly reduced seed yield in Soybeans.

Water stress during the reproductive stage has also been found to influence number and seeds per pod (Sionit and Kramer, 1977; Momen et al., 1979). Water stress at the late reproductive stage accelerated senescence, reduced the seed filling period and resulted in reduction of pod sizes (Costable and Hern, 1978; De Sousa et al., 1997; Brevedan and Egli, 2003). Korte et al. (1983) after comparing three irrigations on eight cultivars of Soybean concluded that a single irrigation during pod elongation was the most beneficial to Soybeans because it increased the number of seeds per plant and irrigation at seed enlargement increases the seed weight. Irrigation of Soybeans at any stage did not significantly increase yield or only slightly increased the yield above that of un-irrigated treatment if the rainfall is sufficient to supply the water requirement (Hunk et al., 1986); irrigation water was not enough (Martin et al., 1979), or if the un-irrigated plant extracted sufficient water from shallow groundwater (Reicosky and Deaton, 1979).

CWP of Soybean can be increased by eliminating irrigation at the vegetative stage when evapotranspiration is predominantly by evaporation from the soil. There are variations in the level of reduction in the yields from one place to the other where DI is practiced. Environmental and soil factors determine the level of soil water evaporation and availability of water in the soil for plant use. Therefore, there is a need to carry out a comprehensive assessment of the impact of DI on the yields of crops in a particular location before implementing it as a policy program. This assessment will be used in convincing farmers and other stakeholder on the benefits that may be derived from such an approach

A parameter for assessing the effect of DI on crop yield is called the crop response factor (k_y). It is a measure of sensitivity of a crop to DI (Doorenbos and Kassam, 1979). Crop response factors have been found to vary from one crop to the other, cultivar and stage of growth when DI was imposed and duration of the DI, irrigation method and management. A value of k_y greater than one indicates that the expected relative decrease in the yields for a given evapotranspiration deficit is proportionally greater than the evapotranspiration deficit (Kirda et al., 2002). The level of accuracy of the crop response factor depends on having sufficient range and data for yield and evapotranspiration and assumes a linear relationship over the data (Doorenbos and Kassam, 1979). The results of the field experiments conducted in the dry seasons are presented in the sections that follow.

8.1   Environmental conditions during the irrigation seasons

Table 8.1 shows the meteorological data in the dry seasons. The average seasonal maximum and minimum air temperatures in the 2013 irrigation season were 36.8 and 20.5 °C respectively, while the average seasonal (maximum and minimum relative humidity are 94.7 and 27.1% respectively. The average seasonal maximum and mean global SRs were 926 and 165 W m^{-2} (Appendix C). The average seasonal maximum and minimum air temperatures in the 2013/2014 irrigation season were 34.6 and 18.8°C respectively. The average seasonal maximum and minimum relative humidity were 100 and 21.7% respectively while the seasonal maximum and mean global solar radiations were 859 and 169 W m^{-2} respectively. This shows that the first season was warmer than the second season. The seasonal effective rainfall depths in the 2013 and 2013/2014 irrigation seasons were 261 and 50 mm, respectively (Table 8.1).

Table 8.2 shows the results of the analyses carried out on the soil samples collected at the experimental site. The average bulk densities in the upper and lower 50 cm were 1.53 and 1.59 g cm^{-3} respectively. The upper 50 cm was richer in OM (average of 0.98%) compared with the lower 50 cm (average of 0.38%). The Ph ranged from 6.1 to 6.4 in the upper 50 cm of the soil profile, which was higher than that of the lower 50 cm (5.3 to 6.0). The lower 50 cm of the soil was higher in phosphorus and iron, average of 30.7 ppm and 1.33 c mol kg^{-1} respectively, than the upper 50 cm, average of 24.2 and 1.23 c mol kg^{-1} respectively. However, the average total N, Na and K in the upper and lower 50 cm of the soil profile were uniform. The upper 50 cm was sandy loamy while the lower 50 cm contained more clay than the upper 50 cm profile. The lay out of the experimental field is shown in Figure 8.1.

Table 8.1. Meteorological data measured at the weather station located near the experimental fields in the dry seasons (Standard deviations in parenthesis)

Year/Month		Temperature (°C)			Relative humidity (%)			Global solar radiation (Wm⁻²)		Rainfall (mm)
		Max	Min	Mean	Max	Min	Mean	Max	Mean	Mean
2013	Feb	41.0	18.0	27.5 (3.7)	94.3	10.1	66.0 (18.6)	904	161(234)	55.3
	Mar	34.5	21.3	27.2 (3.4)	94.4	42.4	76.4 (14.0)	810	128(219)	32.3
	Apr	34.8	21.7	25.8 (3.7)	94.5	40.4	78.5 (13.7)	1003	190 (266)	44.9
	May	37.0	20.8	26.1 (2.7)	95.6	15.6	81.5 (12.9)	985	181 (245)	129
2013/14	Nov	33.5	20.5	26.3 (2.8)	100	37.9	87.2 (22.3)	959	180(265)	-
	Dec	33.1	16.7	25.9(3.3)	100	20.3	78.6(23.5)	837	179(250)	50
	Jan	35.4	18.1	26.4(3.2)	100	15.1	81.3(25.2)	841	152(219)	-
	Feb	36.3	19.7	27.5(3.7)	100	13.5	68.8(25.4)	798	166(229)	-

Table 8.2. Physical and chemical properties of the soil during the irrigation seasons

	00-10	10-20	20-30	30-40	40-50	50-60	60-70	70-80	80-90	90-100
Sand	74	76	64	78	64	54	58	56	64	62
Clay	16	14	26	12	6	36	32	28	12	24
Silt	10	10	10	10	30	10	10	16	24	14
Texture class*	Sandy loam	Sandy loam	Sandy clay loam	Sandy loam	Sandy loam	Sandy clay	Sandy clay loam	Sandy clay loam	Sandy loam	Sandy clay loam
BD	1.41	1.57	1.54	1.57	1.57	1.58	1.56	1.58	1.58	1.66
FC	0.18	0.18	0.26	0.28	0.18	0.21	0.31	0.31	0.31	0.26
PWP	0.08	0.08	0.17	0.09	0.08	0.09	0.21	0.21	0.21	0.16
TAW	0.10	0.10	0.10	0.10	0.11	0.12	0.10	0.10	0.10	0.10
OM	1.41	1.14	0.74	0.74	0.87	0.34	0.47	0.40	0.27	0.40
Total N	0.63	0.46	0.53	0.56	0.53	0.49	0.46	0.55	0.51	0.56
P	21.2	20.8	27.5	25.2	26.1	23.5	27.8	32.8	30.0	39.3
Na^+	0.29	0.27	0.24	0.25	0.21	0.22	0.22	0.26	0.25	0.25
K^+	0.09	0.07	0.08	0.08	0.08	0.07	0.08	0.06	0.07	0.08
Ca^{2+}	28.8	27.8	30.3	32.0	31.7	32.4	29.2	28.7	30.0	27.5
Mg^{2+}	43.7	38.8	45.9	44.8	26.5	22.0	43.7	42.3	31.5	30.7
Fe^{2+}	1.19	0.96	1.06	1.67	1.27	1.33	1.31	0.99	1.32	1.68
Ph	6.4	6.4	6.3	6.2	6.1	5.9	6.0	5.7	5.5	5.3

Sand, clay, Silt, O.M (organic matter) and Total Nitrogen (N) in (%), BD: Bulk density (g cm⁻³); Na^+, K^+, Ca^{2+} and Mg^{2+} in (cmolkg⁻¹), P = phosphorus (ppm), FC (Field capacity) and PWP (permanent wilting point) in (m³ m⁻³), Ph (H₂0). *USDA classification

Figure 8.1. Application of water to Soybeans by using in-line drip irrigation

8.2   Biometrics and growth parameters

8.2.1   Leaf area index and dry matter

Table 8.3 shows the averages of the LAIs for all treatments at flowering, pod initiation, seed filling and maturity stages when water application was skipped in both seasons (sub-section 6.3.1). In the first irrigation season, the highest LAIs were obtained in the reference treatment (Treatment 1 (T_{1111})) that was irrigated every week throughout the growing period while Treatment 4 (T_{1101}) had the minimum LAIs at flowering, pod initiation, seed filling and at maturity. Variability in the soil fertility could be responsible for lower LAIs of T_{1101} even before the treatment started. Peak LAIs in the reference Treatment 1 (T_{1111}) were 33, 36, 41 and 50% higher than in the Treatments 2, 5, 3 and 4 respectively. Treatment 1 (T_{1111}) had the highest LAI because it was not allowed to undergo soil moisture stress. There was no significant difference (p > = 0.05) between the LAIs for Treatments 2 (T_{0111}) where irrigation was skipped during flowering and Treatment 3 (T_{1011}) and 4 (T_{1101}) where irrigation was skipped at pod initiation and commencement of maturity respectively (Appendix M). Treatment 4 (T_{1101}) had the lowest LAI because of the long period of water stress imposed on it.

For the 2013/2014 season, Treatment 1 (T_{1111}) had the highest LAIs at flowering, pod initiation, seed filling and commencement of maturity. The LAIs in these stages in the second season were lower. The crop had the highest LAI in the first season during seed filling (86 DAP). Skipping of irrigation every other week during seed filling in Treatment 4 (T_{1101}) led to significant reduction in the LAI. This was because it was done during mid season, which lasted for 36 days (sub-sections 6.3.5 and 6.3.6), unlike in Treatment 2 (T_{0111}) where skipping of irrigation took place for only seven days because of shortness of the flowering. The LAIs of Treatment 2, 3 and 4 were not significantly different (p > 0.05) at pod initiation and seed filling because reduction in canopy caused by water stress during flowering was compensated when it was adequately irrigated later in the season. However, in the second season, the crop reached peak LAIs during the flowering stage (Table 8.3). Reduction in LAIs when irrigation was skipped during flowering and pod initiation was due to reduction in shoot growth, leaf expansion and shoot display (Blum, 1996), but in Treatment 4, it was due to accelerated senescence of the leaves (Muchow et al., 1986; David et al., 1998; Lecoeur and Guilioni, 1998).

Table 8.3. Leaf area index (m^2 m^{-2}) at flowering, pod initiation, seed filling and maturity during the irrigation seasons

	Treatment Label	FL (R1) 49 DAP	PI (R3) 63 DAP	PF (R6) (86 DAP)	MT (R7-R8) (109 DAP)
2013	1.T_{1111}	3.83 ± 0.40^a	5.46 ± 0.31^a	7.10 ± 0.26^a	2.19 ± 0.12^a
	2.T_{0111}	3.13 ± 0.20^{ab}	4.63 ± 0.17^b	4.79 ± 0.41^b	0.64 ± 0.25^c
	3.T_{1011}	2.81 ± 0.20^b	4.89 ± 0.31^{ab}	4.17 ± 0.81^{bc}	0.64 ± 0.22^c
	4.T_{1101}	2.70 ± 0.47^b	4.53 ± 0.18^b	3.61 ± 0.20^c	0.47 ± 0.09^c
	5.T_{1110}	3.25 ± 0.69^{ab}	4.84 ± 0.57^{ab}	4.53 ± 0.21^b	1.43 ± 0.18^b
2013/2014	1.T_{1111}	3.44 ± 0.27^a	2.61 ± 0.11^a	1.29 ± 0.30^a	1.01 ± 0.01^a
	2.T_{0111}	2.36 ± 0.40^{ab}	2.13 ± 0.06^b	1.01 ± 0.04^a	0.72 ± 0.01^d
	3.T_{1011}	2.96 ± 0.09^{abc}	2.28 ± 0.24^{ab}	1.12 ± 0.05^a	0.79 ± 0.01^b
	4.T_{1101}	3.15 ± 0.40^{ab}	2.25 ± 0.21^{ab}	0.95 ± 0.01^a	0.44 ± 0.01^e
	5.T_{1110}	2.68 ± 0.00^{bc}	2.51 ± 0.14^{ab}	1.16 ± 0.13^a	0.75 ± 0.01^c

FL - Flowering, PI - Pod initiation, SF - Seed filling; MT - maturity. Values of the mean ± SD from four replicates. Means of the LAIs with the same letter are not significantly different at 5% (p > 0.05) level based on Duncan multiple comparison of means.

Pooled over the seasons and across the water regimes, peak plant heights and their corresponding LAIs were not significantly correlated (p > 0.05, SEE = 1.84). This indicates that increments in plant heights did not produce correspondent increment in LAIs (Figure 8.2a). Under full irrigation (FI), the plant attained a peak height of 54 cm. LAI and number of leaves were significantly correlated (p < 0.05, SEE = 0.72). Treatment 1 had the highest number of leaves of 248 and 105 in the 2013 and 2013/2014 irrigation seasons (Figure 8.2b). Number of seeds per plant (SEE = 46.8), number of seeds per pod (SEE = 0.10) and LAIs were polynomially and significantly correlated in the two seasons (p < 0.05) with the maximum of 572 and 276 seeds per plant stand in 2013 and 2013/2014 irrigation seasons respectively (Figure 8.2 c and d).

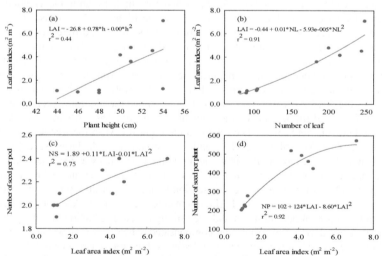

Figure 8.2. Relationship between LAIs during seed filling and (a) plant height; (b) number of leaves; (c) number of seeds per pod; (d) number of pods per plant stand in both irrigation seasons

Table 8.4 shows the DAB in the two seasons. There were seasonal variabilities in the effects of water stress on dry matter. Compared with Treatment 1 (T_{1111}) that was irrigated every other week, the dry matter in Treatment 2 (T_{0111}) reduced by an average of 11.7% (p > 0.05) due to water stress imposed on it during flowering (Appendix N). Water stress at the pod initiation reduced dry matter significantly (p < 0.05) by an average of 21.7% compared with the dry matter in Treatment 1 (T_{1111}). Similarly, water stress during seed filling and commencement of maturity reduced dry matter by seasonal averages of 15% (p > 0.05) and 28% (p < 0.05) respectively. Dry matter reached the peak during the seed filling in the 2013/2014 irrigation season unlike in the 2013 irrigation season when it reached the peak at maturity. Higher LAIs in Treatment 1 (T_{1111}) where water was applied throughout the season resulted into formation of higher canopy and interception of the incident SR by the crop and produced higher dry matter. Figure 8.3 shows the relationships between the pooled data of plant height at harvest, seasonal transpiration, number of pods per stand (NP), number of seeds per pod (NS) and dry matter for the two seasons.

Table 8.4. Dry matter (g m^{-2}) accumulation during the irrigation seasons

	Treatment Label	FL (R1) 49 DAP	PI (R3) 63 DAP	PF (R6) 86 DAP	MT (R7-R8) 109 DAP
2013	1.T_{1111}	127±0.22^a	221±0.35^a	424±0.32^a	578±0.03^a
	2.T_{0111}	114±0.37^{ab}	213±0.47^{ab}	294±0.04^c	503±0.38^a
	3.T_{1011}	68±0.17^b	169±0.08^b	233±0.13^c	439±0.62^{ab}
	4.T_{1101}	99±0.26^{ab}	231±0.14^a	361±0.16^{ab}	458±1.32^{ab}
	5.T_{1110}	94±0.32^{ab}	165±0.06^b	352±0.36^{ab}	359±0.13^b
2013/2014	1.T_{1111}	80.3±22.2^a	112±7.40^a	219±45.9^a	135±1.41^a
	2.T_{0111}	69.8±8.57^a	70.9±28.4^a	148±26.1^a	113±2.12^b
	3.T_{1011}	63.7±21.0^a	65.4±11.9^a	135±10.9^a	113±1.41^b
	4.T_{1101}	58.2±19.0^a	80.6±22.0^a	186±7.41a	108±0.71^c
	5.T_{1110}	74.5±9.50^a	103±4.87^a	158±81.0^a	110±0.71^{bc}

Values are means ± SD from three replicates. Means of DAB with the same letter are not significantly different at 5% (p > 0.05) level based on Duncan multiple comparison of means

Plant heights and DAB were not significantly correlated (p > 0.05, SEE = 168) and increase in plant height did not produce corresponding increment in dry matter. This was due to water stress at different stages, which reduced elongation of the plant. Dry matter at harvest and seasonal transpiration were significantly correlated (p < 0.05, SEE = 111). This means that the production of dry matter depended largely on the amount of transpired water and that the model is responsible for 74% of the dry matter produced (Figure 8.3b). The equation in Figure 8.3b shows that an increment of seasonal transpiration by 50 mm would produce an increment of 23 g m^{-2} of dry matter. The number of pods per plant (SEE = 61.2), number of seeds per pod (SEE = 122) and dry matter were significantly correlated (p < 0.05) and increments in both parameters resulted in corresponding increments in dry matter across the water regimes (Figure 8.3 c and d).

Table 8.5 shows the number of seeds per plant and pods after harvest and variabilities in the number of leaves and plant height during seed filling in the two irrigation seasons. The treatments had the maximum number of leaves and reached the peak height during the seed filling in both season. DI reduced the number of leaves by 19.4, 13.3, 25.8 and 1.6% for Treatment 2 (T_{0111}), Treatment 3(T_{1011}), Treatment 4 (T_{1101}) and Treatment 5 (T_{1110}) respectively in the 2013 irrigation season. Similarly, number of leaves reduced by 22.9, 13.3, 14.3 and 1.0% in Treatments 2, 3, 4 and 5 respectively in the 2013/2014 irrigation season. The plant height was also reduced by 5.6, 7.4, 4.6 and 1.9% in Treatments 2, 3, 4 and 5 respectively in the 2013 season and the corresponding reductions in the 2013/2014 season were 14.8, 18.5, 11.1 and 11.1% respectively. This shows that the effects of DI on the canopy was higher than its effects on the height of the plant. DI reduced the plant height significantly during pod initiation in the two seasons. It reduced the number of seeds per plant by 26.3, 13.8, 9.4 and 19.4% for the Treatments 2, 3, 4 and 5 respectively in the first season while in the second season, it reduced by 24.3, 17.8, 27.2 and 20.3% respectively.

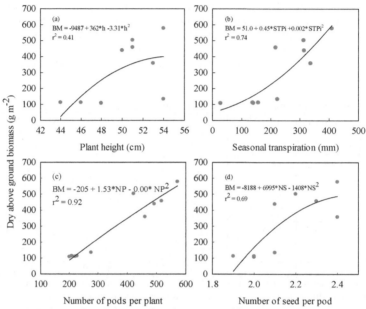

Figure 8.3. Relationships between dry above ground biomass at harvest and (a) plant height at harvest; (b) seasonal transpiration; (c) number pods per plant; (d) number of seeds per pod in both irrigation seasons

Table 8.5. Number of seeds per plant and pods, maximum number of leaves in the mid season and maximum plant heights at harvest

	Treatment label	Number of leaves	Height of plant (cm)	Number of seeds per plant	Number of seeds per pod
2013	1. T_{1111}	248 ± 1^a	54 ± 1^a	572 ± 6^a	2.4 ± 0.1^a
	2. T_{0111}	200 ± 1^d	51 ± 1^a	422 ± 6^b	2.2 ± 0.1^{ab}
	3. T_{1011}	215 ± 1^c	50 ± 1^b	493 ± 5^{ab}	2.1 ± 0.3^b
	4. T_{1101}	184 ± 1^e	51 ± 0^a	518 ± 6^{ab}	2.3 ± 0.1^{ab}
	5. T_{1110}	244 ± 3^b	53 ± 2^a	461 ± 5^b	2.4 ± 0.1^a
2013/2014	1. T_{1111}	105 ± 5^a	54 ± 1^a	276 ± 3^a	2.1 ± 0.1^a
	2. T_{0111}	81 ± 8^a	46 ± 1^a	209 ± 4^a	2.0 ± 0.2^a
	3. T_{1011}	91 ± 8^a	44 ± 1^b	227 ± 6^a	1.9 ± 1.1^a
	4. T_{1101}	90 ± 20^a	48 ± 1^a	201 ± 6^a	2.0 ± 0.1^a
	5. T_{1110}	104 ± 8^a	48 ± 1^a	220 ± 3^a	2.0 ± 0.2^a

Values are means ± SD from three replicates. Means with the same letter are not significantly different at 5% (p > 0.05) level based on Duncan multiple comparison of means.

DI during flowering reduced the number of seeds per plant more than during seed filling. This was possibly due to reduction in the flower production and abortion of flowers (Winket et al., 1997; Egli, 2005). Treatment 1 (T_{1111}) had the highest number of seeds as expected because it remobilized more dry matter from the vegetative parts to the seeds during reproduction than all other treatments (Andriani et al., 1991). Reduction in the seed number due to long periods of water stress during seed filling was also due to reduction in photosynthetic capacity of the leaves and assimilate remobilization (Wardlaw, 1990). This is similar to the findings of Andriani et al. (1991). Although, number of seeds per pod and LAIs were positively correlated, the number of seeds per pod was not affected by DI in the second irrigation season. DI only significantly reduced the number of seeds per pod by 12.5% in 2013 in Treatment 3 (T_{1011}) (Table 8.5).

Table 8.6 shows the growth stages and the seasonal evapotranspiration in the two seasons. Variations in the daily water use at each stage of the crop resulted in differences in the cumulative crop water use. The minimum amount of water was used during the initial stage of the crop while the peak amount was used during the mid season characterised by flowering, pod initiation and filling. Treatment 1 had the maximum seasonal water use in the two seasons. This was expected because it was irrigated weekly throughout the growing period. The seasonal water use of the crop reduced in the treatments where DI was applied. For instance, the crop water use during the mid season in Treatment 1 was 8.8, 15.8, 19.0 and 20.9% higher than the water used for Treatments 5, 3, 4 and 2 respectively. Similarly, in 2013/2014 season, crop water use in Treatment 1 was 6.3, 7.9, 5.6 and 43.8% higher than for Treatments 2, 3, 5 and 4 respectively.

Table 8.6. Growth stages and their actual evapotranspiration (mm), seasonal evapotranspiration (mm) and number of irrigations in the 2013 and 2013/2014 irrigation seasons. The duration of the growth stage in each year is in parenthesis

	Treatment	Establishment (0 - 25)	Vegetative (26 - 58)	Mid season (59 - 100)	Late season (101 - 112)	Seasonal crop water use (mm)	No of weekly irrigation	No of days irrigation was skipped
2013	1.T_{1111}	35	173	273	42	523	9	-
	2.T_{0111}	35	170	216	42	463	8	7
	3.T_{1011}	35	173	230	42	480	8	7
	4.T_{1101}	35	173	221	36	465	8	14
	5.T_{1110}	35	173	249	38	495	8	7
		0-25	26-57	58-100	101-109			
2013/2014	1.T_{1111}	29	130	304	44	507	16	-
	2.T_{0111}	29	123	285	44	481	15	7
	3.T_{1011}	29	101	280	44	454	15	7
	4.T_{1101}	29	130	171	34	364	13	21
	5.T_{1110}	29	129	287	22	467	15	7

8.2.2 *Relationship between accumulation of dry matter, seed yield and seasonal water use*

The relationship between yields and water use is of importance to farmers and other stakeholders in the irrigation industry because it is used in evaluating the effects of yield loss at different levels of water use, especially under limited water supply. Figure 8.4 shows the linear relationship between seed yields, dry matter and crop water use in both seasons. The linear relationships indicate that there seed yields and dry matter increased with increase in the applied water under DI practice in the study area. The linear equations relating the seed yields, DAB and SWU in both seasons are as follows:

$$YD = 11.1 \times SWU_i - 3390 \qquad\qquad r^2 = 0.40 \qquad\qquad (88)$$

$$DM = 17.4 \times SWU_i - 5570 \qquad\qquad r^2 = 0.20 \qquad\qquad (89)$$

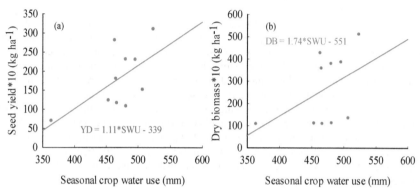

Figure 8.4. Seasonal crop water use, seed yield and accumulation of dry matter in the dry seasons

Equation 88 implies that a threshold of about 306 mm of water is required to initiate seed yield and that an increment of 50 mm of SWU will produce an increment of 555 kg ha^{-1}. Similarly, Equation 89 implies that a threshold of about 321 mm of water is required to initiate an increase in dry matter and that a dry matter of about 870 kg ha^{-1} will be obtained for every increment of 50 mm of SWU. These dry matter and seed yields are significant and fall within the range in Figure 8.4. The presence of residual moisture in the soil cannot be used to produce such high yields of Soybean in the study area. The linear model reported by Nielsen (1990) (Yield (kg ha^{-01}) = 65.3 × SWU (mm) - 1130) predicted similar yield that is about 15% higher than the yield predicted in Equation 86. Exponential model of the yields and seasonal water use (Y (kg ha^{-1}) = 45.4e$^{0.01 \times SWU(mm)}$) (r^2 = 0.48) in this study implies that residual soil moisture will produce a yield threshold of about 45 kg ha^{-1} and thereafter seasonal increment of 50 mm will produce yield at an exponential rate.

8.2.3   Relationship between yield decrease and decrease in evapotranspiration

Figure 8.5 shows the relationship between decrease in seed yields and seasonal deficit evapotranspiration for both irrigation seasons. The regression equation obtained by using the popular water production function of Doorenbos and Kassam (1979) is:

$$\left(1 - \frac{Y_a}{Y_m}\right) = 2.24 \times \left(1 - \frac{SWU_a}{SWU_m}\right) \qquad (90)$$

where:

SWU_a = seasonal crop water use in the treatments where irrigation was skipped every other week during flowering, pod initiation, seed filling and commencement of maturity (mm)

SWU_m = seasonal crop water use in T_{1111} (mm)

Y_a = yield obtained from the treatment where irrigation was skipped every other week during flowering, pod initiation, seed filling and maturity stages and water was applied throughout the season (t ha^{-1})

Y_m = yields obtained from T_{1111} (t ha^{-1})

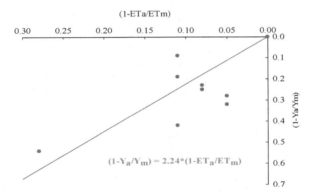

Figure 8.5. Relationship between the decrease in relative yield and deficit in seasonal relative evapotranspiration for the two irrigation seasons

The severity of the moisture stress and the proportionality of the yield decrease relative to DI is called crop response factor, which is the slope of the regression equation (Figure 8.5). The seasonal k_y of 2.24 is higher than 0.85 for Soybean under DI (Doorenbus and Kassam, 1979). This implies that the moisture stress imposed on the crop was severe and the rate of decrease in yield is proportionally higher than the relative deficit evapotranspiration.

Reduction in the seed yields of Soybean is inevitable under DI (Eck et al., 1987; Nielsen, 1990; De Costa and Shanmugathasan, 2002; Karam et al., 2005). In the 2013 irrigation season, yield reductions were 9.3, 25.4, 41.8, 25.7% (p < 0.05) for Treatments 2, 3, 4 and 5 respectively (Appendix O). Similarly, in the 2013/2014 irrigation season, yield reductions in the Treatments 2, 3, 4 and 5 were 28.3, 18.4, 53.9 and 23.0% respectively. Average seasonal reductions in the yields were 18.8 and 21.9% (p > 0.05) for the Treatment 2 and 3 where irrigation was skipped every other week during flowering and pod initiation (Table 8.6). Similarly, average seasonal and significant reductions were 47.9 (p < 0.05) and 24.4% (p > 0.05) for Treatment 3 (T_{1101}) and Treatment 4 (T_{1110}), where irrigation was skipped during seed filling and commencement of maturity. This implies that withholding water during the seed filling and commencement of maturity in Soybeans could lead to significant reduction in the seed yields. Water stress during seed filling was more severe for Soybean as it reduced yield of the crop by half. This occurred because of the reduction in the periods of seed filling and enough assimilates (C and N) were not remobilised to the seeds during long periods of water stress (Blum, 1996; de Souza et al., 1997). Findings in this study support the conclusion that water stress during R3, R5 or R6 resulted in substantial yield reductions compared with FI, with the greatest reduction with water stress at the R6 stage (Westgate and Peterson, 1993; de Souza et al., 1997; Dogan et al., 2007a). The decrease in the seed yield in this study is attributed to the reduction in the number of pods (seeds) per plant stand and is in line with the results of Momen et al. (1979). The yield reduction when irrigation was skipped at the commencement of maturity is not significantly different from the fully irrigated treatment.

The yields of Soybean in this study varied from 1.81 t ha^{-1} (T_{1101}) to 3.11 t ha^{-1} (T_{1111}) in 2013 and 0.70 t ha^{-1} (T_{1101}) to 1.52 t ha^{-1} (T_{1111}) in the 2013/2014 irrigation season (Table 8.7). Compared with deficit irrigated treatments, the seed yields of the fully irrigated treatment were higher than DI treatments in both seasons and were similar to Sincik et al. (2008). T test at 95% confidence limit shows that the average seasonal seed yields of the treatments are significantly different (p < 0.05). The seed yields in this study especially for fully and DI treatments were similar to those previously reported. For instance, yields of fully irrigated treatments in this study fall within the range of 3.6 to 3.7 t ha^{-1} reported by Dogan et al. (2007b) for fully irrigated Soybean but were higher than the average seed yields under different DI treatments. Similarly, the yield range in this study is similar to 2.16 to 3.93 t ha^{-1} and 1.98 to 3.59 t ha^{-1} in 2005 and 2006 irrigation seasons respectively (Candogan et al., 2013); 2.3 to 3.5 t ha^{-1} under different DI (Karam et al., 2005) and 2.07 to 3.76 t ha^{-1} (Sincik et al., 2008).

Table 8.7. Seasonal evaporation, transpiration, crop water use and seed yields in the two irrigation seasons

	Treatment label	SEP$_i$ (mm)	STP$_i$ (mm)	SET$_i$ (mm)	Δ ET	Δ Yield	Yield (t ha^{-1})
2013	1.T$_{1111}$	114±3.05^d	409±4.89^a	523±2.00^a	0.00	0.00	3.11±0.77^a
	2.T$_{0111}$	150±4.54^c	313±5.92^c	463±1.57^d	0.11	0.09	2.82±0.29^a
	3.T$_{1011}$	165±4.13^{ab}	315±5.89^c	480±2.00^c	0.08	0.25	2.32±0.26^{ab}
	4.T$_{1101}$	160±4.16^{ab}	217±4.51^d	465±0.50^d	0.11	0.42	1.81±0.40^b
	5.T$_{1110}$	158±4.98^b	337±6.01^b	495±1.20^b	0.05	0.32	2.31±0.35^{ab}
2014	1.T$_{1111}$	284±7.78^b	223±4.32^a	507±2.12^a	0.00	0.00	1.52±0.28^a
	2.T$_{0111}$	324±13.0^a	157±2.34^b	481±2.12^b	0.05	0.28	1.09±0.30^{ab}
	3.T$_{1011}$	316±3.56^a	138±4.23^c	454±3.54^d	0.11	0.19	1.24±0.29^a
	4.T$_{1101}$	336±4.24^a	28±3.11^d	364±0.71^e	0.28	0.54	0.70±0.03^b
	5.T$_{1110}$	327±8.49^a	140±1.24^c	467±1.41^c	0.08	0.23	1.17±0.31^{ab}

Values are means± SD from three replicates. Means of the yields, SEP$_i$, STP$_i$ and SWU$_i$ with the same letter are not significantly different at 5% (p > 0.05) level based on Duncan multiple comparison of means

8.3   Soil water balance

8.3.1   Seasonal soil evaporation (SEP$_i$), transpiration (STP$_i$), and water use (SWU$_i$)

The components of the soil water balance considered in this study are soil water evaporation, plant transpiration, and seasonal crop water use. Runoff and deep percolation were very small and therefore were considered negligible. Table 8.7 shows the seasonal evaporation (SEP$_i$), seasonal transpiration (STP$_i$), seasonal crop water use (SET$_i$), in the two seasons. Irrigation was applied at intervals of seven days in each season. Treatments 1 (T_{1111}) and 5 (T_{1110}) were irrigated the same number of times but the former received the highest irrigation depths in the 2013 irrigation season. The lengths of each stage and rainfall event that occurred during the crop cycle were responsible for the differences in the total amount of water applied to each treatment. Significant differences ($p < 0.05$) were observed in the SEP_i, STP_i, and SET_i in the 2013 irrigation season, indicating that there is variability in the water used under DI in the stated treatments. Treatment 1 (T_{1111}) had the peak STP_i and (SET_i) while Treatment 4 (T_{1101}) had the minimum STP_i in both seasons. Higher STP_i and SET_i in the control treatment was expected because it was irrigated more often than any other treatment during the growing season unlike Treatment 4 where irrigation was skipped every other week during the mid season that had the longest duration in the growth cycle of the crop. For instance, in the first season, irrigation was skipped for a total period of 14 days whereas in the second season, it was skipped for 21 days. Evaporation reduced significantly by 30.9, 9.1, 3.0 and 4.2% in Treatments 1, 2, 4 and 5 respectively in the 2013 irrigation season compared with Treatment 3 (T_{1011}). Similarly, in the 2013/2014 irrigation season, evaporation reduced by 15.5, 3.60, 6.00, and 2.7% in Treatments 1, 2, 3 and 5 respectively compared with Treatment 4 (T_{1101}). Treatment 1 (T_{1111}) received the highest amount of water that favoured denser canopy and higher LAIs than other treatments during the growing seasons (Table 8.4). Evaporation was 21.8, 31.9, 32.4, 34.4 and 34.4% of the seasonal crop water use in the Treatments 1, 5, 2, 3 and 4 respectively in the 2013 irrigation season. In the 2013/2014 season, evaporation was 56.0, 67.4, 69.6, 70.0 and 92.3% of the seasonal water use. Evaporation was more pronounced in the 2013/2014 irrigation season, as it constituted a seasonal average of 71% of the seasonal crop water use unlike the 2013 irrigation season where the comparative seasonal average was 31% of the crop water use. Higher proportion of the crop water use was partitioned towards non-productive evaporation and this was responsible for the lower yields in the second irrigation season (Table 8.6). Number of leaves and leaves area indices in Treatment 4 reduced significantly due to the extended period of water stress and thereafter reduced the photosynthetic capacity.

STP$_i$ in the 2013 irrigation season reduced significantly by 23.5, 23.0, 46.9 and 17.6% ($p < 0.05$) in Treatments 2, 3, 4 and 5 respectively due to water stress. Similarly, in the 2013/2014 irrigation season, transpiration reduced significantly by 29.6, 38.1, 87.4 and 37.2% ($p < 0.05$) in Treatments 2, 3, 4 and 5 respectively. Average seasonal transpiration in the 2013 and 2013/2014 irrigation seasons constituted about 70 and 30% respectively of the SWU_i. Pooled over the years, LAIs at seed filling and seed yield (Table 8.7) were significantly correlated by using a polynomial model ($p < 0.05$, SEE = 25.2). It implies that the highest or potential seed yield of 350×10 kg ha^{-1} was obtainable at LAI of 11.5 m^2 m^{-2}. However, this could not be reached as a result of water stress and environmental conditions during the irrigation seasons. The peak LAI of 7.10 m^2 m^{-2} was obtained in the first irrigation season (Figure 8.6a). Pooled over the years, a linear model fitted to STP_i and seed yields were highly significant (Y (kg ha^{-1}) = $0.67 \times STP_i$ (mm) + 29.5; $r^2 = 0.92$, $p < 0.05$). This means that 92% of the variability in the seed yield can be explained by STP_i and that for every increment of 10 mm in STP_i, seed yield will increase by 6.7 kg ha^{-1}. Reduction in the transpiration for the treatments under deficit water application was responsible for the lower yields compared with Treatment 1 (T_{1111}) that was fully irrigated in the two seasons. This suggests that increasing the amount of transpired water will have significant increment in the seed yields of Soybean in the study area. High correlation between seed yield and STP_i is similar to findings of Purcell et al. (2007). LAIs were significantly correlated, with seasonal transpiration ($p < 0.05$, SEE 53.6) (Figure 8.6b). This indicates that LAIs account for 79% of the variability in STP_i. The linear model gave a positive but weak correlation ($r^2 = 0.25$, $p > 0.05$, SEE = 1.98) between LAIs at seed filling and seasonal crop water use.

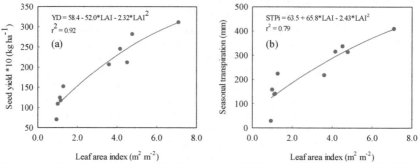

Figure 8.6. Relationship between LAIs during seed filling and (a) seed yield; (b) STP_i in the two seasons

The SWU_i of Soybean and other crops under irrigated conditions vary from one area and season to the other (Lamm et al. 2010). Seasonal cop water use of 463 to 523 mm and 364 to 507 mm for both irrigation seasons fall within the crop water use reported in literature. Seasonal evapotranspiration of 554 to 721 mm for Soybeans was reported by Gercek et al. (2009) and 513 to 1261 mm by Lamm et al. (2010). Similarly, Candogan et al. (2013) reported seasonal evapotranspiration of between 394 and 802 mm in 2005 and 351 to 841 mm in 2006 under different levels of DI conditions. Dogan et al. (2007b) reported seasonal crop water use ranging from 574 mm to 619 mm for fully irrigated conditions.

8.4   Water productivity and irrigation water productivity (IWP)

WP ranged from 3.89 kg ha mm^{-1}for Treatment 4 (T_{1101}) to 6.09 kg ha mm^{-1} for Treatment 2 (T_{0111}) while IWP for the same treatments ranged from 8.9 kg ha mm^{-1} in Treatment 4 (T_{1110}) to 14.0 kg ha mm^{-1} in Treatment 2 (T_{0111}) in the 2013 irrigation season (Table 8.8). The WP in this study is within the range of 4.4 to 5.1 kg ha^{-1} mm^{-1} as reported by Liu et al. (2013) for Soybean. Treatment 1 (T_{1111}) that was irrigated throughout the season did not produce the highest IWP, while Treatment 2 (T_{0111}) in which irrigation was skipped every other week during flowering gave the highest IWP in the first season. IWP_{seed} for DI during flowering (T_{0111}) was 15% higher than that of Treatment 1 (T_{1111}). This trend supports Howell et al. (1990), who stated that while maximum WP tends to occur at maximum crop water use, maximum IWP usually occurs at crop water use less than the maximum. Based on this, Howell et al. (1990) suggested that irrigating to achieve the maximum crop yield and crop water use would not be the most efficient use of irrigation water. The results obtained in this study show that IWP of Soybeans can be increased if irrigation is skipped during flowering. Treatment 2 (T_{0111}) had the highest WP and IWP while Treatment 4 (T_{1101}) had the minimum in the 2013 irrigation season. However, in the 2013/2014 irrigation season, Treatment 1 (T_{1111}) had the peak WP_{seed} and IWP_{seed} and Treatment 4 had the minimum WP and IWP. For instance, in the 2013 irrigation season, WP_{seed} for Treatment 2 (T_{1011}) was 2.3, 16.1, 23.5, and 36.1% higher than the WP for the Treatments 1, 3, 5 and 4 respectively. In the same season, IWP for Treatment 2 was 15, 20, 29.3 and 36.4% higher than for the Treatments 1, 3, 5 and 4 respectively. In the 2013/2014 irrigation season, however, WP for Treatment 1 (T_{1111}) was 8.7, 16.3, 24.7 and 35.7% higher than the WP for the Treatments 3, 5, 2 and 4 respectively. Similarly, IWP_{seed} was 7.2, 15.4,

24.1 and 32.5% higher than for the Treatments 3, 5, 2 and 4 respectively. The result indicates that under water limited conditions, skipping of irrigation every other week during flowering, can be used to increase WP and IWP of Soybeans. However, skipping of irrigation during the seed filling stage in Treatment 4 (T_{1101}) will greatly reduce the seed yields of the crop. SWU and WP over the water regimes and years were not significantly related by using linear model (WP = 0.02*SWU - 4.83; r^2 = 0.26, p = 0.13). Increase in SWU by 10 mm will only increase the WP by 0.18 kg ha^{-1} mm^{-1}. Pooled over the seasons, both WP, IWP and seed yield ware linearly and significantly correlated (Y (kg ha^{-1}) = 51.1*WP (kg ha^{-1} mm^{-1}) - 13.8, r^2 = 0.98, p < 0.05, SEE = 13.2 kg ha^{-1}); (Y = 16.4*IWP +66.3, r^2 = 0.90, p < 0.05, SEE = 27.6 kg ha^{-1}). These indicate that seed yields of the crop may not be compromised by increasing the WP and IWP. The WP fall within 4.58 to 5.58 kg ha^{-1} mm^{-1} as reported by Sincik et al. (2008)

Table 8.8. Water productivity, irrigation water productivity and harvest indices for full and deficit irrigation

Treatment label	2013 irrigation season			2013/2014 irrigation season		
	WP_{seed} (kg ha^{-1} mm^{-1})	IWP_{seed} (kg ha^{-1} mm^{-1})	HI (%)	WP_{seed} (kg ha^{-1} mm^{-1})	IWP_{seed} (kg ha^{-1} mm^{-1})	HI (%)
1.T_{1111}	5.95	11.9	61.3±2.9[abc]	3.00	3.32	63.9±7.8[a]
2.T_{0111}	6.09	14.0	65.9±1.6[a]	2.26	2.52	56.1±4.6[ab]
3.T_{1011}	5.11	11.2	62.4±5.5[ab]	2.74	3.08	55.4±3.2[ab]
4.T_{1101}	3.89	8.9	56.0±3.0[c]	1.93	2.24	43.2±12.5[b]
5.T_{1110}	4.66	9.9	59.6±1.3[bc]	2.51	2.81	47.6±4.5[ab]

Agronomic and environmental factors such as method of irrigation, environmental conditions and potential yields of the crop influence WP and IWP. The results obtained in this study show that WP and IWP for a high yielding variety such as (TGX 1448 2E) can be improved by using in-line drip irrigation. Similar experiments could be performed for sprinkler and surface irrigation in the study area to ascertain their performances.

8.5   Water productivity and harvest index

Figure 8.7 shows the linear model between WP, IWP and HI in the two irrigation seasons. HIs for Treatments 4 (T_{1101}) significantly reduced by 15.1% (p < 0.05) during the seed filling and 5.35, and 9.60% for Treatments 3 (T_{1011}) and 5 (T_{1110}) respectively in the 2013 irrigation season. Similarly, HI of Treatment 4 significantly reduced by 32.4% during seed filling and 12.2, 13.3 and 15.0% in Treatments 2, 3 and 5 respectively in the 2013/2014 irrigation season. Substantial reduction in HI in Treatment 4 was due to the fall in transpiration because of consecutive depletion of the moisture in the root zone, which aborted fruits set, reduced fruit filling and hence reduced the yield. This trend shows that water stress during seed filling can reduce significantly the HI of Soybeans under DI, which is in agreement with de Souza et al. (1997). Pooled over the seasons and water regimes, HI and WP are positively correlated (r^2 = 0.53, p < 0.05, SEE = 1.12; r^2 = 0.78 for 2013 and r^2 = 0.66 for 2013/2014). Similarly, HI and IWP are positively correlated (r^2 = 0.44, p < 0.05, SEE = 3.36; r^2 = 0.90 for 2013 and r^2 = 0.62 for 2013/2014) season. The minimum HIs obtainable for the cultivar under investigation were 33.2 and 40.5% for the seasonal WP and IWP respectively. Improvement in the WPs and IWPs in this study was due to improved HIs in the treatments subjected to water stress at the stated growth stages. Based on the data, it can be inferred that the cultivar TGX 1448 2^E had efficient canopy in producing seeds. Results of this study are in agreement with Neyshabouri and Harfield (1986) and Westgate et al. (2004) who suggested that WP of Soybeans could be improved by increasing its HI.

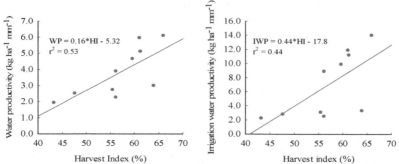

Figure 8.7. Water productivity as a function of harvest index of TGX 1448 2E Soybeans grown under drip irrigated conditions with five treatments

8.6   Effects of deficit irrigation on yield components

The yield components at harvest considered are the seeds (S_d), dry chaffs (C_f) and stems (S_t).There is variability in the proportions of the dry chaffs and stems in each season (Figure 8.8). For full irrigation in the first season, chaff and stem constituted 24 and 15.2% respectively when the seed yield was 3.11 t ha^{-1}. In Treatment 4 (T_{1101}) where the seed yield was 1.81 t ha^{-1}, the proportions of the dry chaffs and stems were 23.4 and 16.2% respectively. In the 2013/2014 irrigation season, when the seed yield for full irrigation was 1.52 t ha^{-1}, the proportions of the chaffs and stem were 33.1 and 34.6% respectively. With the minimum seed yield of 0.77 t ha^{-1} for Treatment 4, chaffs and stems constituted 39.4 and 55% respectively. Pooled over the years, seed and chaff were significantly correlated by using a linear model ($C_f = 0.14 + 0.33S_d$; $r^2 = 0.86$, p < 0.05, SEE = 0.10 t ha^{-1}). However, chaffs and stems ($r^2 = 0.65$, p = 0.07); seed and chaff ($r^2 = 0.27$, p = 0.57, SEE = 0.11t ha^{-1}) were not correlated by using polynomial models.

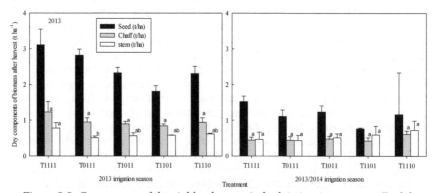

Figure 8.8. Components of the yield at harvest in both irrigation seasons. Each bar represents the average from three replicates and standard error after the components have been oven dried except the seed

8.7   Economic evaluation

The economic evaluation of the full and DI approaches under land and water limiting conditions is shown in Table 8.9. During decision-making, factors such as land productivity, WP, IWP and revenues for each irrigation strategy need to be considered. In this study, average yields, WP, IWP and revenues for each treatment were compared. Average seasonal WP for full irrigation was 4.48 kg ha^{-1} mm^{-1} while it was 2.91 kg ha^{-1} mm^{-1} for DI during seed filling. However, DI at flowering had an average maximum IWP of 8.26 kg ha^{-1} mm^{-1} and DI during seed filling had an average minimum IWP of 5.57 kg ha^{-1} mm^{-1}. High WP and IWP indices are of little interest if they are not associated with acceptable seed yield, production cost and total revenue (Oweis and Hachum, 2004). Irrigation water was most productive by skipping it every other week during flowering than at any other stage (Table 8.8). Interestingly this was associated with a relatively good average seed yield of 1.96 t ha^{-1} compared with 2.32 t ha^{-1} for full irrigation.

Table 8.9. Economic evaluation of the use of the drip method in cultivating Soybeans under full and deficit irrigation conditions in Ile-Ife

Treatment (1)	Yield (t ha^{-1}) (2)	Total cost of production ha^{-1} (US$)×10^3 (3)		Price per ton (US$) (4)	Total revenue (US$)×10^3 (2*4) (5)	Loss (US$)×10^3 (3-5) = (6)		Economic productivity US$ ha^{-1} mm^{-1} (7)
		(a)	(b)			(a)	(b)	
1. T$_{1111}$	2.32	12.4	6.01	541	1.30	11.1	4.71	2.42
2. T$_{0111}$	1.96	11.9	5.93	541	1.06	10.8	4.87	2.26
3. T$_{1011}$	1.78	11.9	5.93	541	0.96	10.8	4.97	1.83
4. T$_{1101}$	1.26	11.0	5.70	541	0.68	10.3	5.02	1.57
5. T$_{1110}$	1.74	11.9	5.93	541	0.94	11.0	4.99	1.94

Scenario (a) accounted for the cost of water used and other fixed costs in the total cost of production while scenario (b) does not

8.7.1   Land limited conditions

Correct application of DI requires a thorough evaluation of the economic impact of the yield reduction caused by water stress (Sepaskhah and Akbari, 2005). Under land limiting conditions (without considering opportunity cost of irrigation water), the optimum water application strategy is that which maximises net return per unit of land (Ali, 2010). The total revenue increased with increase in seasonal crop water use (Figure 8.9). The total revenues for T$_{0111}$, T$_{1011}$, T$_{1110}$ and T$_{1101}$ reduced by 18.5, 26.2, 27.7 and 47.7% respectively compared with T$_{1111}$. For scenario (a), the financial loss for skipping irrigation during maturity, flowering, pod filling and seed filling was 0.90, 2.70, 2.70 and 7.21% respectively compared with FI. Considering scenario (b), the financial loss for skipping irrigation during maturity, pod initiation, flowering and FI compared with skipping of irrigation during seed filling was 0.60, 1.00, 2.99 and 6.18% respectively (Table 8.9). Although no financial benefit was obtained for the two scenarios, FI had the highest total revenue and the minimum loss under scenario (b) and is considered the best irrigation option under land limited conditions.

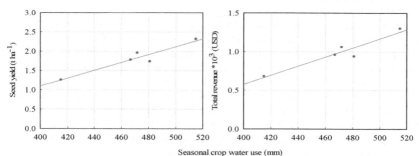

Figure 8.9. Seasonal crop water use versus (a) seed yield and (b) total revenues under land limiting conditions

8.7.2 Water limited conditions

WP$_{economic}$ increased with increasing amount of water applied to the crop (Table 8.9). Average seasonal WP$_{economic}$ of 2.42 US\$ ha^{-1} mm^{-1} (0.24 US\$ m^{-3}) under FI was higher than those of the Treatments 2, 5, 3 and 4 by 6.7, 19.9, 24.3 and 35.0% respectively. This indicates that economic output of water increases with the amount of water evapotranspired. The cost of production increased with increasing number of irrigations (Figure 8.10a). Although the number of irrigations accounted for 56% of the variability in the WP$_{economic}$, there was no significantly correlation (Figure 8.10b). In the first scenario the cost of drip lines and the accessories constituted between 36.9 to 41.3% of the total cost, while water constituted between 54.2 to 59.3% (Appendix P). The fixed cost constituted between 37.5 to 41.9% of the total cost of production. The skipping of irrigation for 7 days in the Treatments 2, 3 and 5 reduced the cost of production by 4.0% while in Treatment 4 T$_{1101}$, water stress for 21days reduced the cost of production by 11.3% compared with T$_{1111}$. However, for scenario (b) the cost of production reduced by 1.3% for skipping of irrigation during flowering, pod initiation and maturity while skipping of irrigation during seed filling reduced the cost of irrigation by 5.2% compared with FI. In the second scenario, the cost of drip line and the accessories only constituted between 75.6 to 76.7% of the total cost of production while the fixed cost were between 76.9 to 80.9% of the total cost.

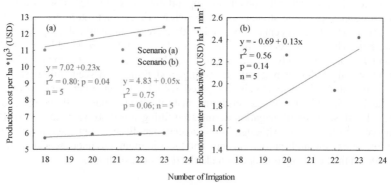

Figure 8.10. Number of irrigations versus (a) production cost for the two scenarios and (b) economic water productivity under water limiting conditions

Shortage of irrigation water is often a constraint in crop production, especially in the dry seasons. In such water limiting conditions, the water saved by DI in a piece of land can be used to irrigate additional land and thereby increasing farm income (English et al., 1990; Geerts and Raes, 2009). Skipping of irrigation for 7 days during flowering, pod initiation and maturity conserved about 8.23, 12.2 and 11.4 litres of water per m^2 respectively. DI for 21 days during seed filling conserved 14.1 litres m^{-2} (141 mm ha^{-1}). It constituted 22.1% of average SWU_i for FI to produce 2.32 t ha^{-1} and 23.7 to 27.5% for DI to produce 1.26 to 1.96 t ha^{-1}. The water conserved could be used for increasing land productivity for Soybeans or cultivating other crops such as vegetables in addition to Soybeans during dry seasons in the study area. Under such conditions, Treatments 2 to 5 are the appropriate irrigation strategies.

Despite higher yields in Treatment 1, the maximum revenue of US$ 1300 could not provide any financial return under both scenarios let alone Treatment 4 that received minimum amount of irrigation in the two seasons (Table 8.9). This clearly shows that the use of drip irrigation in the cultivation of the crops is not financially sustainable for a peasant farmer in Ile-Ife. The reason is that the cost of importing drip lines from the developed countries are too high. Sometimes, the cost of shipping is higher than the cost of irrigation equipment. A peasant farmer may only benefit from the use of drip lines after several years of continuous cultivation of the crop and adequate maintenance of the facility during cultivation and off-season or if the entire fixed cost of production is financed by the government or a donor agency. The production of drip lines locally by using less expensive and durable materials could reduce the total cost of production. The financial benefits at the end of a cropping seasons depends on strategies used in reducing the cost of production and the available price of the crop in the market.

8.8   Conclusion

The results show that DI of Soybean reduced canopy cover, leaf area index, number of seeds per plants, dry matter, grain yield and crop water use. DI reduced transpiration and thereafter the dry matter and seed yields. Duration of the flowering, pod initiation, seed filling, maturity and the total number of days that irrigation was skipped also contributed to the severity of the effects of DI on LAI, dry matter, crop water use and yield of Soybean. DI at flowering, pod initiation, seed filling and commencement of maturity reduced seed yields of the crop. Although DI at flowering and pod initiation affected LAI, compensation was made after subsequent water application during the season and the effects on the dry matter and seed yield were minimal. Subjection of Soybeans to water stress for consecutive 7 days during flowering and total of 21 days during seed filling did not significantly reduce the number of seeds per pod. Due to the long period of seed filling, DI during seed filling affected LAI and dry matter such that further application of water during the short period of maturity could not compensate for the reduction and thereby resulted in significant reduction in the seed yield. Peak WPs were obtained under full irrigation conditions in the two irrigation seasons, IWP may be improved to or above the level of full irrigation if water application will be skipped at flowering and pod initiation. In a situation where water is very scarce due to dry spells in the study area and there is a need to spread DI over the growing season WP and IWP may be increased by avoiding irrigation during pod initiation and commencement of maturity. The water conserved during DI could be used to cultivate other crops and thereby increase land productivity and opportunity cost in the study area. Under land limiting conditions, full irrigation is hereby recommended. Under water limiting conditions, weekly skipping of irrigation during flowering and total elimination of irrigation during maturity are recommended. The outcomes of the economic analysis under land and water limiting conditions provide information for policy makers at basin scales for formulating improved and efficient water management plans in regions with similar weather conditions. The results will be beneficial in adopting DI in a manner that will improve WP and increase $WP_{economic}$ at local and international scales.

9. Modelling response of the growth and yield of Soybeans to full and deficit irrigation by using the AquaCrop model

9.1 Introduction

Agriculture is a key occupation and a means of livelihood in Nigeria, especially among the rural dwellers. Although effort is being made to harness groundwater and streams for crop production, rainfall is still the major source of water for farming in many of the agro-ecological zones in Nigeria (Table 2.1). Rainfall in many parts of the country is often erratic and unreliable. For instance, in the south-western (Sub-humid) part of Nigeria, no effective rainfall was recorded in 2012 until the month of June in Ile-Ife, which was contrary to the usual occurrence of rainfall in February. The torrential nature of rainfall in Ile-Ife causes high runoff and waterlogging, which reduce land productivity. Despite relatively higher rainfall in Ile-Ife, dry spells occur at any time even during the rainy season, which triggers severe water stress and reduces yield of the crops. Dry spells are being experienced often in the recent times and may be attributed to effects of climate change in the region. Future climate projections indicate that stress due to water and high temperature could limit crop productivity (Vorosmarty et al., 2000). In the previous years, traditional agriculture placed emphasis on increasing total production because land and water were not limiting. Due to scarcity of water in many regions, deficit irrigation (DI) is now given attention and several researches have been carried out on the responses of crops to deficit application of water (Nautiyal et al., 2002; Henggeler et al., 2002; Payero et al., 2005).

A model is defined as a simplification of reality. It is an approximation of complex reality that was designed to give better understanding of the real world system (Ali, 2011). Crop growth is a complex system, which consists of varying parameters interacting with one another. By specifying a set of environmental conditions, crop models can be used to predict growth, development and seed yields of a crop (Monteith, 1996). Three crop growth models were identified as reported in (Abendinpour, 2012) and these are (i) carbon driven (ii) radiation driven and (iii) water driven. The water driven models assume that a linear relationship exists between the rate of growth of biomass and amount of water transpired through a water productivity parameter (Steduto and Albrizio, 2005). An advantage of water driven models over radiation driven models is that the former can normalize WP parameters for the ET_0 and atmospheric CO_2 and therefore can be used or applied to locations with varying environmental conditions (Steduto et al., 2007).

Over the years, many sophisticated crop growth models have been developed and used under different agro-ecological conditions for modelling growth of annual and perennial crops. These models include WOFOST for yield of Wheat (Ma et al., 2013), eight different models for simulation of yield of winter Wheat (Palosuo et al., 2011) and CROPGRO for Soybean (Sau et al., 1999; Dogan et al., 2007b). Most of these models require high data input, which is difficult to obtain for some crops, especially in the developing countries and highly skilled personnel for their calibration and validation. Many of the models are plant or cultivar specific e.g Water Accounting Rice Model (WARM) and CERES-Rice or ORYZA and designed to be used under specific environments. These requirements therefore limit their application to regions where model input data are available and the environments that they were designed and calibrated for. The existing models consist of complex components and systems such as LAI, water potential, which are mostly understood by scientists. However, the targeted end users such as water users associations and policy makers find them difficult to follow and use (Fereres, 2011). In order to address these challenges and constraints, and to achieve an optimum balance between accuracy, simplicity, and robustness, FAO developed the crop model called AquaCrop (Raes et al., 2012). This model has been used in modelling growth of crops in different parts of the world. For instance, yield and biomass of Maize under varying irrigation and Nitrogen regimes in Semi Arid environments of India was calibrated and validated by using the model (Abedinpour et al., 2012). Under all irrigation and Nitrogen regimes, the prediction error in simulating grain yield ranged from a minimum of 0.47 to 5.9% and a maximum of 4.36 to 11.1% for biomass. The model prediction error for the WP varied from 2.35 to 27.5%. Hsaio et al. (2009) used 6 years data on Maize at Davis Experimental Station CA, The AquaCrop model simulated final above ground biomass within 10% of the measured value for at least 8 of the 13 treatments and in at least four of the considered cases, the simulated yields and biomasses were within 5% of the measured values. The Wilmott's Index of agreement for 11 out of 13 cases was ≥ 0.98 for CC and ≥ 0.97 for biomass. Wheat yield (grain) was modelled by using the model for data obtained from five experimental sites in Canada with r^2, d and RMSE of 0.66, 0.99 and 743 kg ha^{-1} respectively. Observed and modelled soil water content produced a r^2 of 0.90, d of 0.99 and RMSE of 49 mm and a Mean Absolute Error (MAE) of 40 mm (Mkhabela and Bullock, 2012). Araya et al. (2010) evaluated the performance of the model in simulating the yields and biomass of Barley for different dates and sites and discovered that the model is valid. Steduto et al. (2009b) conducted experiments for seven years on Soybean in India and stated that 1 in 6 predictions of biomass and grain yields was outside the 5% deviation of the observed values by using the model. They further stated that the model showed sensitivity to initial soil moisture conditions and advised that validation of Soybean parameters in the semi-arid tropics of India was important. According to Evett and Tolk (2009), experimentation cannot address all scenarios but appropriate modelling can fill in the gap. Four models performed well in simulating yields and CWPs of crops such as cotton (*Gossypium hirsutum* L.), Maize (*Zea mays* L.), Quinoa (*Chenopodium quinoa* Willd.), and Sunflower (*Helianthus annuus* L.) in the North and South America, Europe, and the Middle East under well watered conditions but misestimated CWP under DI. However, separate measurements of E and T, which could have furnished information on the poor performances of these models under water stress conditions, were not mentioned in all the experiments and findings. They advised the use of models that separate E from T_r in further studies on modelling of crop yields so that the effects of soil conservation such as mulching and bunding or the combination on reducing evaporation can be quantified.

Todorovic et al. (2009) compared the performances of the models AquaCrop, CropSyst and WOFOST in simulating the growth of Sunflower and discovered that several parameters in the modules did not have substantial influence on the results. He recommended the use of the AquaCrop model, which requires lesser input data and has a higher accuracy in simulating yields, biomass and CWP of crops for management purposes in future studies. Therefore, in this study the AquaCrop model has been calibrated and validated for Soybeans cultivated at TRFOAU, Ile-Ife under full and DI conditions. The site descriptions were given in the in sections 6.1 and 8.1 while, field lay out, experimental treatments and agronomic measurement during the dry seasons were presented in section 6.3.

9.2   Input data requirement of the AquaCrop model

Simulation of yield and other parameters in the AquaCrop model requires input data of the environment and crop components in the three panels on the user interface of the model. The environment and crop panel consist of climatic data; crop data, soil data, and the management component consist of field and mode of water application (rainfed or irrigation management). Selection and adjustment of these parameters can be done for varieties of crops for specific environments during calibration and simulation.

9.2.1   Environmental conditions

The weather data required for the AquaCrop model are daily minimum and maximum temperature, minimum and maximum relative humidity, wind speed, SR, relative sunshine hours, rainfall, reference crop evapotranspiration (ET_o) and average CO_2 concentration. Reference crop evapotranspiration was determined with the ET_o Calculator (Equation 23) by using daily maximum and minimum temperatures and relative humidity, SR, mean actual vapour pressure, sunshine hours and wind speed at 2 m obtained from the automatic weather station (Vantage Pro2, Davis Instruments Corp., USA) located within a distance of 200 m from the experimental site. The graphical illustrations of the daily meteorological data of the study sites during the rainfed and irrigation conditions are shown in the Figures 9.1 to 9.4. Explanation on the variability of meteorological data during the dry seasons is given in section 8.1. Details of the irrigation treatments, schedule of water application and the method used in measuring actual crop evapotranspiration during the dry seasons were in given section 6.3.

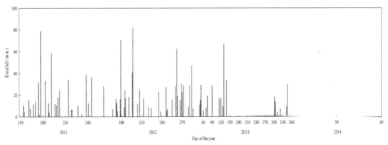

Figure 9.1. Daily rainfall at the experimental fields from April 2011 to February 2014

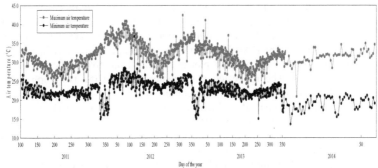

Figure 9.2. Daily temperature at the experimental fields from April 2011 to February 2014

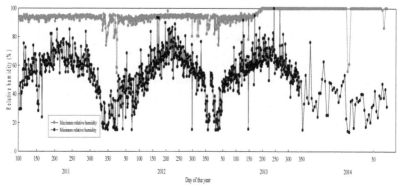

Figure 9.3. Daily relative humidity at the experimental fields from April 2011 to February 2014

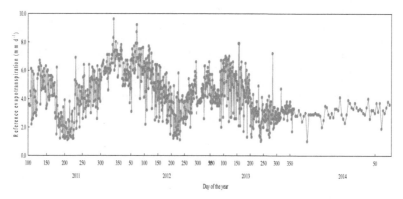

Figure 9.4. Daily reference evapotranspiration at the experimental fields from April 2011 to February 2014

9.2.2   Crop parameters

The development of canopy was measured in terms of the phenologic growth stages, LAI, CC, root length, and DAB. During the crop cycle, the date of emergence of the crop, CC_x, time from planting to flowering, duration of the flowering, initial and maximum canopy cover, the date of commencement of the senescence and maturity at late stage were recorded. The DAB in each treatment was determined by harvesting one plant (0.179 m^2) at the centre of each plot at random from two replicates only at an interval of seven days. The harvested plant materials from the two replicates were combined together and oven dried at a temperature of 70 ^{0}C for 48 hours and the weight was recorded after a constant value was obtained (section 6.3). Conservative parameters, which can be adapted to different locations and cultivars, were taken from AquaCrop Reference Manual and include CGC and CDG; the crop coefficient for transpiration at full CC etc. (Table 9.1) (Raes et al., 2009). The CC for each treatment was determined with Equation 70 (sub-section 6.2.3). In the AquaCrop model, the sensitivity and severity of moisture depletion in the soil is determined by the upper and lower thresholds and the shape of the response curve are the parameters for each type of stress. For instance, the upper threshold determines when the stress begins, while the lower threshold is the point at which the physiological process in plants completely ceases. The shape factor used in the AquaCrop model describes the amplitude of the stresses, which affect the yield of the crop. A shape factor of zero means that the crop shows highest sensitivity to water stress and greater than zero is an indicator of less sensitiveness to water stress. The water stress is categorized into expansion stress (Ks_{exp}), stomata closure stress (Ks_{sto}) and senescence stress coefficients (Ks_{sen}). These coefficients were calibrated by using the experimental data to obtain an excellent match between the measured data on the field and simulated output of the model.

Table 9.1. Conservative parameters used in simulating the response of Soybeans (Raes et al., 2012)

Symbol	Description of parameters	Values/range	Unit or meaning
T_{base}	Base temperature	5.0	°C
T_{upper}	Cut-off temperature	30.0	°C
CC_o	Canopy cover per seedling at 90% emergence	5.00	cm^{-2}
CGC	Canopy growth coefficient		Increase in CC relative to existing CC per day
$K_c\ Tr_x$	Crop coefficient for transpiration at CC = 100%	1.10	Full canopy transpiration relative to ET_o
	Decline in crop coefficient after reaching CC_x	0.30%	Decline per day due to leaf aging
CDC	Canopy decline coefficient at senescence		Decrease in CC relative to CC per GDD
WP*	Water productivity normalized for ET_o and CO_2	15.0	$g\ m^{-2}$ (biomass)
$K_{s\ exp}$ upper	Soil water depletion threshold for canopy expansion - Upper threshold	0.15	As fraction of TAW, above this leaf growth is inhibited
$K_{s\ exp}$ lower	Soil water depletion threshold for canopy expansion - Lower threshold	0.65	Leaf growth stops completely as this p
	Leaf growth stress coefficient curve shape	3.0	Moderately convex shape
P_{sto}	Soil water depletion threshold for stomatal control - Upper threshold	0.5	Above this stomata begins to close
	Stomata stress coefficient curve shape	3.0	Highly convex curve
P_{sen}	Soil water depletion threshold for canopy senescence - Upper threshold	0.7	Above this early canopy senescence begins
	Shape factor for Water stress coefficient for canopy senescence	3.0	Convex curve
	Coefficient describing positive impact of restricted vegetative growth during yield formation on HI	None	HI increased by inhibition of leaf growth at anthesis
	Coefficient describing negative impact of stomata closure during yield formation on HI	Strong	HI reduced by inhibition of stomata at anthesis
	Allowable maximum increase (%) of specified HI	10	-

9.2.3 Soil parameters

Soil parameters are important in the operation of the AquaCrop model. The soil data used as input parameters are: five soil horizons, textural class of the soil at the experimental field, field capacity (θ_{FC}) of the textural class of soil, permanent wilting point (θ_{PWP}), saturated hydraulic conductivity (K_{sat}), and the volumetric water content of the soil at saturation. The hydraulic properties of the soil in Table 9.2 and the corresponding default values of the (K_{sat}), and volumetric water content (θ_{sat}) of the soil at saturation in the AquaCrop model were used in the simulation. At the experimental site, there was a restrictive layer at 0.80 m that obstructed root elongation. The default curve number (CN) in the model was used to determine the surface runoff from the daily rainfall events that occurred during the experiment.

9.2.4 Irrigation and field management

Irrigation and field management are part of the components of the AquaCrop model. As shown before the irrigation management consisted of full irrigation (T_{1111}), weekly skipping of water application at flowering (T_{0111}), pod initiation (T_{1011}), seed filling (T_{1101}) and commencement of maturity (T_{1110}). Field management components were the fertility levels, mulching to reduce evaporation of water from the soil, furrows and bunds to eliminate surface runoff. In this study, the AquaCrop model was evaluated through calibration and validation to estimate yield, biomass and canopy cover (CC) under varying water applications at specific stages of growth.

9.3   Calibration of the AquaCrop model

Calibration of the AquaCrop model was done by using the data measured in the experimental field during the 2013 irrigation season. The model was used to simulate yield, dry above ground biomass and canopy cover. The simulated output was compared with measured data. Parameters which influence reference variables were adjusted by using a trial and error approach in order to reduce to the barest minimum the numerical difference between the simulated and measured data and to ensure that good matches were obtained between the simulated output and measured data for each treatment.

9.3.1 Calibration of irrigation parameters

In the AquaCrop model there are several options for simulating irrigation. They include determination of net irrigation water requirement and generation of irrigation schedules based on management strategies either for rainfed or irrigated farming. These options are user-specific. The irrigation component contains different irrigation options. These are surface irrigation (i.e. basin, furrow and border), sprinkler irrigation and drip irrigation. The panel contains several percentages of the soil that is wetted under each irrigation, which the user can select from. In the current study, in-line drip irrigation was used to apply water to the crop based on the 30% wetted area option. Irrigation schedules can be developed in the model by using either a fixed time interval based on scheduling or management allowed depletion (MAD) of which 50% was considered as initial condition in the model. In the current study, an irrigation schedule containing the date of application and net water requirement was imputed directly. Provision was not made in the model for input of efficiency of water application or uniformity efficiency.

Table 9.2. Non-conservative parameters used in simulating the response of Soybeans to water in Ile-Ife conditions

Symbol	Description of parameters	Values	Unit or meaning
T_{base}	Base temperature	5.0	°C
T_{upper}	Cut-off temperature	30.0	°C
CC_o	Canopy cover per seedling at 90% emergence	5.00	cm^{-2}
CGC	Canopy growth coefficient	13.2	Increase in CC relative to existing CC per day
CDC	Canopy decline coefficient at senescence	29.1	Decrease in CC relative to CC per day
$K_c Tr_x$	Crop coefficient for transpiration at CC = 100%	1.10	Full canopy transpiration relative to ET_o
	Decline in crop coefficient after reaching CC_x	0.30%	Decline per day due to leaf aging
WP*	Water productivity normalized for ET_o and CO_2	15.0	g m^{-2} (biomass)
$K_{s\ exp}$ upper	Soil water depletion threshold for canopy expansion - Upper threshold	0.14	As fraction of TAW, above this leaf growth is inhibited
$K_{s\ exp}$ lower	Soil water depletion threshold for canopy expansion - Lower threshold	0.65	Leaf growth stops completely as this p
	Leaf growth stress coefficient curve shape	3.0	Moderately convex shape
P_{sto}	Soil water depletion threshold for stomata control - Upper threshold	0.58	Above this stomata begins to close
P_{sen}	Soil water depletion threshold for canopy senescence - Upper threshold	0.70	Above this early canopy senescence begins
	Stomata stress coefficient curve shape	3.0	Highly convex curve
	Shape factor for water stress coefficient for canopy senescence	3.0	Convex curve
	Coefficient describing positive impact of restricted vegetative growth during yield formation on HI	None	HI increased by inhibition of leaf growth at anthesis
	Coefficient describing negative impact of stomata closure during yield formation on HI	Strong	HI reduced by inhibition of stomata at anthesis
	Allowable maximum increase (%) of specified HI	10	-
	Maximum basal crop coefficient K_{cb}	1.15	
	Time from sowing to emergence	10	Day
	Time from sowing to start of flowering	43	Day
	Duration of flowering	12	Day
	Time from sowing to start of senescence	97	Day
	Time from sowing to maturity	112	Day
	Duration of building up of the harvest index (HI)	69	Day
	Minimum effective rooting depth	0.3	M
	Maximum effective rooting depth	0.8	M

9.3.2   Calibration of field management practices

The field management file of the AquaCrop model contains default data on fertility of soil, crop residue and soil surface practices. For each treatment, the same level of soil fertility (near optimal) was considered because the same amount of fertilizer was applied to all treatments. A curve number of 62, which is the default in the model, was used for the study area. Accumulation of water on the soil surface (surface ponding) was not observed on the field during the experiment and mulches were not applied and therefore not considered in the field management module. Mulches were not found on the field before planting, operation and after the crop was harvested, plant residues were not left on the field. Therefore, they were not considered in the off season module.

9.3.3 Calibration of the crop parameters

The crop file in the AquaCrop model has five major components and associated dynamic responses. These are phenology, foliage canopy, rooting depth, production of biomass and harvestable yield (Steduto et al., 2009a). CC is a key parameter in the model, which determines the amount of water transpired. It is directly related to LAI. There are six parameters that determine the development of canopy. These are CGC, CDC, CC_x, days to emergence, days to senescence and days to full maturity after planting. The rate at which canopy expands is controlled by CGC at which the canopy expands and the CDC controls how fast the canopy reduces at the late stage of a season. Several iterations were done by the trial and error approach in this study by fitting the CGC and CDC until good simulation of the CC was obtained. During the calibration process, the simulated CC was compared with the measured CC at different days after planting. Similarly, the simulated biomass and seed yields were compared with the measured data (Figures 9.5 and 9.6). At the same time the simulated WP was compared with the WP computed data in Table 8.8 for the 2013 irrigation season. The default HI was adjusted to simulate the measured yields in the field. The same procedure was used for all the treatments until the simulated data matched with the measured data.

For full irrigation (T_{1111}), the calibrated CGC and CDC were 13.2 and 29.1%. The CC_x was 96% at near optimal soil fertility level. The days after planting to 90% emergence, maximum canopy, senescence and maturity were 10, 58, 97 and 112 days. The periods of flowering, yield formation and HI constitute the reproductive stages in the AquaCrop model. The model accommodates both the determinant and non-determinant crops. For the determinant crops, canopy expansion ceases after flowering while for the non-determinant crop, canopy expansion continues even after flowering.

Flowering started at 43 DAP and lasted for 12 days. The period spent in building up of the HI was 69 days. The crop attained CC_x during seed filling after which the expansion of CC stopped. The peak effective rooting depth was iteratively set at 0.80 m and was attained during the seed filling stage. The default value of maximum transpiration crop coefficient (Kcb_x) of 1.10 was used for the simulation. The normalized water productivity (WP*) of 15.0 g m^{-2} which falls within the 15 to 20 g m^{-2} acceptable for C3 crops was used in the simulation and adjusted by 15% for yield formation. The C3 plants are plants in which a 3-carbon intermediate acid (phosphoglycerate) is the first stable product during CO_2 fixation (Decoteau, 2005). In the AquaCrop model, the harvestable yield of a crop is the product of HI and biomass accumulated over time. For full irrigation, the calibrated HI was 62% and higher than the default value in the model. This probably was due to differences in the cultivar and genetic characteristics of the crop. The water stress coefficient for canopy expansion was at 0.14 (upper), 0.65 (lower) and a shape factor of 3.0. Similarly the water stress coefficient for stomata closure and early senescence were calibrated at 0.58 (upper) and 0.70 (upper) respectively with the lower threshold set at permanent wilting point (PWP). These stress coefficients were used to calibrate yields of the crop for different treatments and adjustment of the initial cc_o was made where necessary, for instance depending on the time and severity of the stress imposed on the crop until the simulated yield, CC and biomass attained a match.

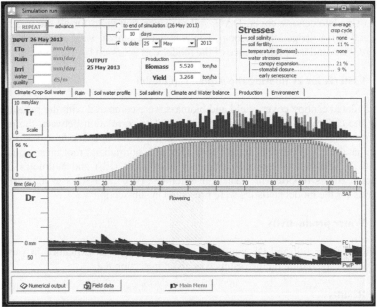

Figure 9.5. Simulated yield of the crop for full irrigation during the calibration of the model by using the measured data of the 2013 irrigation season

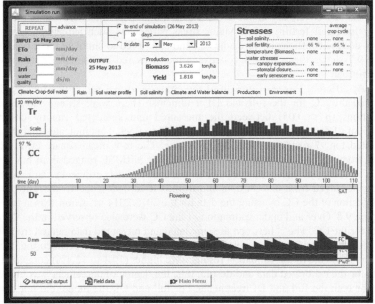

Figure 9.6. Simulated dry biomass of the crop for full irrigation during the calibration of the model by using the 2013 irrigation season data

9.4   Validation of the AquaCrop model

The data obtained in the field in the 2013 season were used for the calibration and validation was done by using the data for 2013/2014. The input data of the model were used to simulate canopy cover, DAG, soil moisture, and yield for all the treatments in 2013 and 2014 irrigation seasons. The simulated output was compared with the measured data and the performance of the model was accessed by using model statistics.

9.5   Criteria for evaluating the AquaCrop model

The simulated canopy cover, yield and biomass were compared with the measured data during the calibration and prediction processes. The correlation between the predicted and measured data was evaluated by using the following models: coefficient of determination (r^2), root mean square error (RMSE), normalized root mean square error (NRMSE), Nash-Sutcliffe model efficiency coefficient (EF) and Wilmott's index of agreement (d) as reported in Nash and Sutcliffe (1970) and Raes et al. (2012). The equations for the model statistics were presented in sub-section 5.2.7.

9.6   Water productivity

WP and IWP for full and DI conditions were determined as explained in sub-section 6.3.4, while explanation on the results was given in section 8.4.

9.7   Results

9.7.1   Canopy cover

The AquaCrop model was calibrated by using data for the 2013 irrigation season to predict canopy cover, dry above ground biomass and soil moisture for full and DI at near optimal soil fertility level. Major stress parameters such as canopy growth and canopy decline coefficients were adjusted and readjusted to simulate the measured CC. Figure 9.7 shows the relationship between the measured and simulated canopy cover with error bar.

At different stages of growth of the crop, the AquaCrop model under and over estimated the CC for the full and DI treatments. There were strong and significant correlations ($p < 0.05$) between the measured and simulated canopy cover with coefficient of determination r^2 ranging from 0.97 to 0.99. This indicates that the model accounted for 97 to 99% of the canopy cover. The root mean square error (RMSE) ranged from 4.3 to 5.8%, which is acceptable. The NRMSE ranged from 6.5 to 9.4%, and is considered excellent for the simulation. Model efficiency and degree of agreement ranged from 0.96 to 0.98 and from 0.96 to 0.99 respectively. The results of the validation of the CC by using the data for the 2013/2014 irrigation season are shown in Figure 9.8. Over and under estimation of the CC were also observed in the validation results for the CC. The r^2 between the simulated and measured data ranged from 0.96 to 0.98 and were significantly correlated ($p < 0.05$). The NRMSEs between the measured and simulated data in the validation results were higher (10.1 to 12.4%) compared to the results for the calibrated data and despite this, the degree of agreement was higher (d = 0.99) for both the full and DI treatments. For full and DI, the model did not delay the date of emergence. This is contrary to the delay in the date of emergence for 2-3 days in the simulation of the CC for Soybeans in India (Steduto et al., 2009b). Minimum NRMSEs were obtained for Treatments 1 (T_{1111}) and 3 (T_{1011}) during calibration while in the validation, Treatments 2 (T_{0111}) and 3 (T_{1101}) had the minimum NRMSEs. The simulation and validation of the CC were satisfactory for both irrigation seasons.

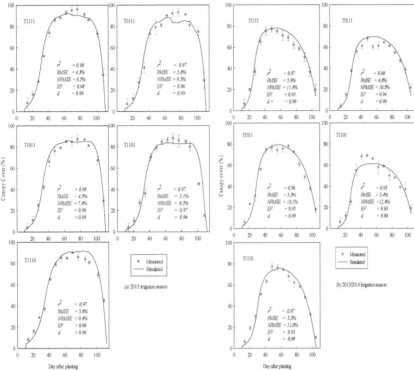

Figure 9.7. Relationship between simulated and measured canopy cover for full and deficit irrigation conditions during (a) calibration in the 2013 and (b) validation in the 2013/2014 irrigation seasons. Each dot with standard error represents the mean of the measured canopy cover from two replicates

9.7.2 Dry above ground biomass

Figure 9.8 shows the simulated and measured dry above ground biomass. The AquaCrop model over estimated dry above ground biomass from emergence until anthesis in the calibrated data for both full and DI. The above ground biomass simulated after anthesis was lower in Treatments 2 (T_{0111}) and 3 (T_{1011}) compared with the similar data for full irrigation (T_{1111}) and Treatments 4 (T_{1101}) and 5 (T_{1110}). There was good agreement between the simulated and dry above ground biomass with r^2 ranging from 0.85 to 0.99, while the RMSEs ranged from 0.14 t ha^{-1} for Treatment 1 (T_{1111}) to 1.09 t ha^{-1} for Treatment 2 (T_{0111}). This implies that the model accounted for 85 to 99% of the simulated data. The NRMSEs ranged from 7.6% for Treatment 1 to 48.1% for Treatment 2. Reduction in the simulated dry above ground biomass after anthesis resulted in higher NRMSEs for Treatments 2. There was a high degree of agreement between the simulated and observed DAG and this was satisfactory. The NRMSEs for the validated data for the dry above ground biomass ranged from 6.3% for Treatment 3 (T_{0111}) to 27.7% for Treatment 2 (T_{0111}) and the degree of agreement between the measured and simulated output were good ($0.95 \leq r^2 \geq 0.99$).

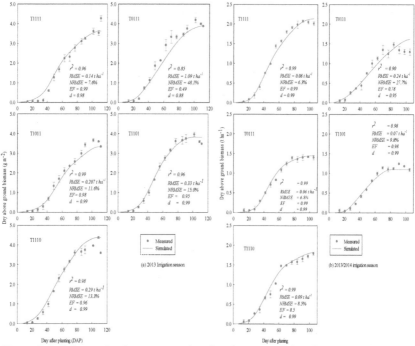

Figure 9.8. Relationship between simulated and measured dry above ground biomasses for full and deficit irrigation during (a) calibration in the 2013 and (b) validation in the 2013/2014 irrigation seasons. Each dot with standard error represents the mean of the measured dry biomass from two replicates

9.7.3 Soil moisture content

Figure 9.9 shows the trend between the simulated and measured soil moisture content for the full and DI conditions. The AquaCrop model slightly overestimated soil moisture content above field capacity (FC) except for Treatment 3 (T_{1011}) during the seed filling and shortly before commencement of maturity and for Treatment 5 (T_{1110}). The model did not simulate moisture content below the permanent wilting point (PWP) in both the full and DI conditions. For instance, the measured soil moisture content was at PWP in Treatment 4 (T_{1101}) where irrigation was skipped for a cumulative period of 21 days and slightly above PWP in Treatment 5 where irrigation was skipped at commencement of maturity. The model underestimated soil moisture content in full irrigation. However, overestimated it for most of the DAP in DI conditions. This is similar to the findings of Mkhabela and Bullock (2012), Farahani et al. (2009) and Hussein et al. (2011) who reported that the model gave good predictions of the wetting and drying cycles due to irrigation events; but tended to consistently overestimate total soil water content, especially in the DI plots.

Zeleke et al. (2011) reported that the AquaCrop model captured the trend between the measured and simulated soil moisture, it tended to overestimate it in most of the time during the cropping periods. A reason can be attributed to the underestimation of the soil moisture content by the model at FC. The AquaCrop model does not allow moisture to remain above FC for consecutive days and similar to other models assumes that soils at saturation drain to FC within a short period. There were high and significant correlations between the measured and simulated soil moisture contents in the upper 0.80 m for full and DI conditions with r^2 ranging from 0.75 for Treatment 2 (T_{0111}) to 0.95 for Treatment 1(T_{1111}). Efficiency coefficient ranged from 0.02 for Treatment 5 (T_{1011}) to 0.81 for treatment 1 (T_{1111}) and d $\geq$ 0.78 for both full and DI conditions. This indicates that the model accounted for 75 to 95% of the variability in the soil moisture content under full and DI of Soybeans in Ile-Ife. The NRMSE ranged from 8.3% for Treatment 1 to 14.9% for Treatment 5 and the d ranged from 0.79 for Treatment 3 to 0.95 for Treatment 1. If the model is to be accessed based on NRMSE, EF and d, it can be concluded that it performed best under full irrigation. Notwithstanding, its performances under DI were satisfactory and commendable.

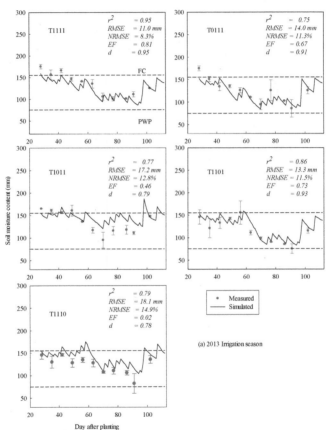

Figure 9.9. Comparison of the simulated and measured soil moisture for full and deficit irrigation at depth of 0 - 0.8 m during the 2013 irrigation season. Each dot represents the mean of the soil moisture from two replicates. Standard error shows the deviation of the soil moisture from the mean value

9.8   Validation results of the AquaCrop model

Table 9.3 shows the comparison of the measured and simulated seed yields, dry above ground biomass and seasonal crop water use and their percentage deviations from the measured values. Both the yield and biomass were simulated adequately. For the seed yields and dry above ground biomass, the excellent performances of the model is demonstrated by the fact that only 1 in 5 predictions was outside the 15% and 20% deviations respectively from the measured values. Simulated seed yields were significantly correlated ($r^2 = 0.99$) with measured values, with RMSE of 0.10 t ha^{-1} (Figure 9.10). Maximum deviations of the simulated yields from the measured values were recorded for Treatment 2 (T_{0111}) and Treatment 4 (T_{1101}) during validation. Similarly, the simulated dry above ground biomass was significantly correlated with the measured data ($r^2 = 0.92$) and RMSE of 0.36 t ha^{-1} (Figure 9.11). Simulated data for Treatment 5 (T_{1110}) and Treatment 2 had the maximum deviations from the measured dry above ground biomass.

Table 9.3. Results of the calibration of seed yield, dry above ground biomass under full and deficit irrigation conditions and percentage deviations of the simulated data from the measured values

	Treatment Label	Yield			Dry above ground biomass			Seasonal crop water use		
		Measured (t ha^{-1})	Simulated (t ha^{-1})	(P_d) ($\pm\%$)	Measured (t ha^{-1})	Simulated (t ha^{-1})	(P_d) ($\pm\%$)	Measured (mm)	Simulated (mm)	(P_d) ($\pm\%$)
2013	1.T_{1111}	3.11	3.27	5.14	4.26	3.62	15.0	523	502	3.97
	2.T_{0111}	2.82	2.87	1.74	3.87	3.89	0.52	463	402	13.2
	3.T_{1011}	2.32	2.42	4.31	3.34	3.59	7.49	480	425	11.5
	4.T_{1101}	1.81	1.86	2.76	3.48	3.82	9.77	465	390	16.1
	5.T_{1110}	2.31	2.40	3.90	3.59	4.37	21.7	495	430	13.1
2013/2014	1.T_{1111}	1.52	1.63	7.24	2.01	2.13	5.97	507	507	0.00
	2.T_{0111}	1.09	1.25	14.7	1.30	1.61	23.8	481	484	0.64
	3.T_{1011}	1.24	1.32	6.45	1.40	1.38	1.43	454	479	5.51
	4.T_{1101}	0.70	0.80	14.3	1.09	1.10	0.93	364	368	1.10
	5.T_{1110}	1.17	1.23	5.13	1.79	1.82	1.68	467	444	4.93

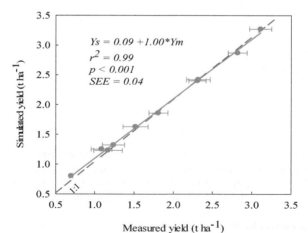

Figure 9.10. Measured and simulated seed yields of Soybeans for the 2013 and 2013/2014 irrigation seasons. Each dot represents the mean seed yield with standard error from three replicates

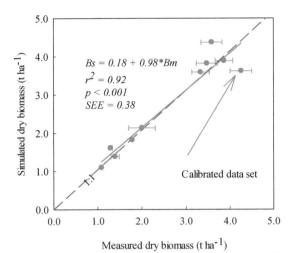

Measured dry biomass (t ha^{-1})

Figure 9.11. Measured and simulated dry above ground biomass of Soybeans for the 2013 and 2013/2014 irrigation seasons. Each dot represents the mean of the dry above ground biomass with standard error from two replicates of the treatments

The simulated water productivity showed high deviations from the measured values in both irrigation seasons (Figure 9.12). It ranged from 18.2% for Treatment 2 (T_{0111}) to 24.5% for Treatment 5 (T_{1110}) in the calibrated data whereas in the validated data, it ranged from 0% for Treatment 5 to 135% for Treatment 2. Despite higher deviations of the simulated data from the measured values, they were significantly correlated with $r^2 = 0.58$ and RMSE of 2.03 kg ha^{-1} mm^{-1}, and the performance of the model was satisfactory.

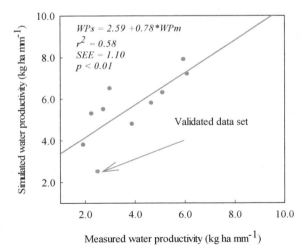

Measured water productivity (kg ha mm^{-1})

Figure 9.12. Measured and simulated water productivity of Soybeans for the 2013 and 2013/2014 irrigation seasons

The AquaCrop model simulated the seasonal ET_a across the irrigation treatments in the 2013 and 2013/2014 irrigation seasons with RMSEs of 58.3 and 15.4 mm respectively (Figure 9.13). The largest and smallest under prediction errors of ET_a were 16.1% for Treatment 4 (T_{1101}) and 3.97% for Treatment 1 (T_{1111}) in the 2013 irrigation season. The under prediction in the full irrigation treatment is incidentally the same as for the simulated deep percolation (data not shown). During the fieldwork, deep percolation was not observed but could not be totally eliminated due to incidental rainfall in a very few days during the seasons. In the 2013/2014 irrigation season, seasonal ET_a was predicted with higher accuracy (data not shown). The largest over prediction error was 5.51% for Treatment 3 (T_{1011}) that corresponded to 25 mm of ET_a in the season. Although the environmental conditions in the two seasons were not too different, the model predicted higher deep percolation in the 2013/2014 irrigation season. The reason for this occurrence is not clear. The trend in the simulated soil moisture content in the 2013/2014 irrigation season was satisfactory and the performance of the model was good. However, very few over prediction of soil moisture above FC did not substantially affect the simulation of the seasonal ET_a, which is a key component of the soil water balance in the model.

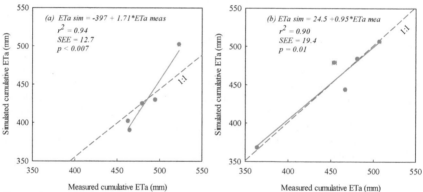

Figure 9.13. Measured and simulated cumulative crop evapotranspiration for Soybeans for (a) 2013 and (b) 2013/2014 irrigation seasons. Each dot represents the mean of the crop evapotranspiration with standard error from two replicates

9.9   Discussion

The model performed excellently in simulating the canopy cover under both full and DI conditions. This is evident by the high r^2 value which ranged from 0.97 to 0.98 with NRMSEs from 5.9 to 6.3% and d = 0.99 for full irrigation. For the DI, r^2 was from 0.93 to 0.98 and d from 0.96 to 0.99, while NRMSE ranged from 7.4 to 12.4%. Variation in the canopy cover could be attributed to different CGCs and CDCs that were iteratively used in the simulation to ensure a good match between the simulated and measured data set. In addition, differences in the time from sowing to flowering, start of senescence and maturity of the crop in the two irrigation seasons were also responsible for the variations. Large reduction in the canopy cover in Treatment 4 (T_{1101}) was due to a long period of drought that subjected the treatment to severe water stress.

The AquaCrop model performed better in simulating dry above ground biomass for full irrigation than for DI conditions. This is because the model efficiency (EF) for full irrigation was 0.99 and d ranged from 0.98 to 0.99 whereas in DI, model efficiency ranged from 0.49 to 0.99. The deviation of the simulated dry above ground biomass from the measured value at harvest ranged from 0.52 to 23.5%. This deviation is likely due to the problems in the calibration data set. This is seen in the deviation of the calibration data set from the 1:1 line (Figure 9.11). Despite these deviations, the RMSEs were considerably lower for both full (0.08 - 0.14 t ha^{-1}) and deficit (0.06 to 1.09 t ha^{-1}) irrigation conditions, which indicates that the model performance is satisfactory. Deviation of the calibrated data set from the 1:1 line for DAB is similar to the findings of Araya et al. (2010).

The seed yields that were simulated by the model compared favourably with the measured data set and deviated slightly from the 1:1 line (Figure 9.10). This is shown by the high r^2 = 0.99, d = 0.99 and low RMSE of 0.10 t ha^{-1}. High performance of the model in simulating the seed yields was due to the use of determined HI$_o$ (Table 8.8) in the calibration and validation of the model and the adjustment made the yield formation to ensure adequate simulation of the yields. Differences in the yields of the same treatment in different seasons could be due to variations in dates of planting and environmental conditions in the seasons.

The model simulated soil moisture adequately in this study. The overestimation of the soil moisture contents above FC at pod initiation and commencement of maturity could result from sampling errors in the measurement of soil moisture during these periods and deviation of the soil input data in the simulation of the soil moisture from the default data set in the model.

The deviation of the simulated WP from the measured value was due to errors encountered in the validation process. This could also be attributed to genetic differences in the cultivars of Soybeans. The cultivar used in this study is different from the default Soybean in the model. Similarly, changes in the environmental conditions in the two seasons could also contribute to deviation of the simulated water productivity from the measured data.

9.10   Conclusion and remarks

The AquaCrop model was calibrated and validated to predict canopy cover, dry above ground biomass, seed yield, crop evapotranspiration, soil moisture content and water productivity of Soybeans in the sub-humid tropics of Nigeria. The simulated data compare adequately with the measured data except for water productivity that was under predicted in the validation data set. The model predicted canopy cover with error statistics of $0.93 \leq E \leq 0.98$ for both full and DI and the degree of agreement d = 0.99 with $4.3 \leq RMSE \leq 5.9$ for full irrigation while for DI, $0.96 \leq d \leq 0.99$ with $5.3 \leq RMSE \leq 5.8$. DAB was predicted with error statistics of $0.08 \leq RMSE \leq 0.14$ t ha^{-1} with $0.98 \leq d \leq 0.99$ for full irrigation while for DI $0.06 \leq RMSE \leq 1.09$ t ha^{-1} with $0.85 \leq d \leq 0.99$ and 1 in every 5 predictions of the above ground biomass was outside the 20% deviation from the measured values.

The yields were predicted with error statistics of RMSE = 0.10 t ha^{-1} and d = 0.99 and 1 in 5 predictions was outside 15% deviation from the measured data. The prediction error statistics for ET$_a$ for both full and DI treatments was $15.4 \leq RMSE \leq 58.3$ in the two seasons. The model over predicted deep percolation also in the validation data set. This indicates that the model needs to be adjusted to ensure better performance. The performances of the AquaCrop model in predicting canopy cover, seed yield and other quantities in this study are commendable and satisfactory.

Specific features such as the use of canopy cover rather than LAI make the model suitable for developing countries like Nigeria where farmers and researchers cannot afford state-of-the-art equipment for measuring LAI. In addition, water productivity normalized for atmospheric demand and carbon dioxide concentration and its focus on water makes it suitable for diverse locations (Stedutos et al., 2009b).

There is no model that is universal in its ability to take into consideration differences in cultivar, environment, weather and management conditions (Farahani et al., 2009; Hsiao et al., 2009). Other cultivars of Soybeans in Nigeria and other agro-climatic environments of the world need to be tested and fine-tuned in the model in order to ascertain the accuracy of the model.

10. Evaluation

In every agronomic plan, the key issue is to improve resource use efficiency in an environmental friendly or sustainable manner. Sustainable practices mean that the needs of the present generation (e.g., yield, resource, and environment integrity) are met without compromising the future in terms of resource degradation or depletion (Matson et al., 1997).

Soybean was cultivated in the rainy seasons under field management practices to conserve water and in the dry seasons under FI and DI. Seasonal crop water use, yield and water productivity under rainfed and irrigated conditions were compared. In this chapter, the resource (land, water and radiation) use efficiency of the crop in rainy and dry seasons are evaluated based on soil water storage (SWS); Cumulative Intercepted Photosynthetically Active Radiation (TIPAR); land and water productivity, financial benefits and economic productivity.

Average seasonal SWSs for the two rainy seasons are in the following order: Conventional practice (NC) < Bunding plus mulching (MLBD) < Tied ridges (TR) < Soil bund (BD) < Mulching (ML) < Tied ridges plus mulching (TRML) < Tied ridges plus bunding (TRBD). This indicates that TRBD had the highest average SWS. Average seasonal TIPAR ranged from 232 MJ m^{-2} for TRBD to 251 MJ m^{-2} for BD (Table 7.7). Average proportion of seasonal transpiration during the rainy season (STP$_r$) in seasonal water use during the rainy seasons (SWU$_r$) ranged from 39.3% for TRML to 43.4% for MLBD while the proportion of seasonal evaporation (SEP$_r$) in SWU$_r$ ranged from 56.6% for MLBD to 60.7% for TRML (Table 7.8). Average seasonal seed yields (land productivity) in the two seasons for TR, TRML, ML, TRBD, BD and MLBD were 24.2, 25.8, 27.4, 33.1, 41.7 and 44.3% respectively higher than that of the conventional practice (Table 7.9). Average seasonal water productivity for the grains (WP$_{seed}$) for TR, ML, TRML, TRBD, BD and MLBD were 14.0, 16.6, 21.0, 24.6, 32.0 and 41.8% higher than that of the conventional practice. This shows that land and water productivity were higher under water conservation practices than under the conventional method. The treatment of the soil and the regular maintenance increased the cost of production where water was conserved compared to the conventional practice. For instance, average seasonal costs of production per hectare for TR, MLBD, TRML and TRBD (2,280 US$) were 28.9% higher than those of ML and NC (1,620 US$), and about 10.1% higher than those of BD (2,050 US$) (Table 7.10). In addition to cost of maintaining the water conservation measures, several other precautionary measures, such as regular weeding, weekly application of insecticides and construction of drains to divert running water away from the fields contributed to the high cost of production. This is normal for a project of this nature to ensure highest accuracy of agronomic practice. Most of these measures are not fully observed among the peasant crop growers. Therefore, their cost of production may reduce a little bit from those presented in this study. Adoption and sustainable practice of these water conservation measures on large scales require investment in medium scale agricultural machinery.

Conservation of water under rainfed conditions increased the economic water productivity of the crop. Minimum economic water productivity for TR (2.64 US$ ha^{-1} mm^{-1}) among the water conservation practices was 14.0% higher than of the conventional practice (Table 7.10). The implication is that for every mm of water that is evapotranspired per hectare under TR and NC, 2.64 and 2.27 US$ respectively are generated to the agricultural sector. MLBD that had the highest land and water productivity also had the highest economic water productivity of 3.90 US$ ha^{-1} mm^{-1}.

Information on economic water productivity of crops is very important for effective water management and accounting at basin scale.

DI led to reduction in land and water productivity of the crop. For instance, average seasonal yield when irrigation was skipped every other week during flowering (T_{0111}), skipped every other week at pod initiation (T_{1011}), skipped every other week during maturity (T_{1110}) and skipped every other week during seed filling (T_{1101}) reduced by 18.7, 30.3, 33.3 and 84.9% respectively compared to without skipping (T_{1111}) (Table 8.9). Yield reduction is inevitable under deficit irrigation. Average seasonal water productivity reduced by 7.2, 14.0, 24.8 and 53.8% for T_{0111}, T_{1011}, T_{1110} and T_{1101} respectively. The highest reduction in land and water productivity under deficit irrigation occurred during seed filling. Increasing the seasonal water use and water productivity led to increase in seed yield of the crop. The average seasonal costs of production are similar for all irrigation scenarios. They ranged from 11,000 US$ ha^{-1} for T_{1101} to 12,400 US$ ha^{-1} for T_{1111} if the cost of water is included in the total cost of production (Table 8.9). If the cost of water is removed from the total cost of production, it ranged from 5,700 US$ ha^{-1} for deficit irrigation during seed filling to 6,100 US$ ha^{-1} for full irrigation. Reduction in the cost of production under DI compared to full irrigation was due to the costs of water and equipment under scenarios (a) and (b) respectively. Generally, the seasonal cost of pumping under scenario (a) can be reduced by using alternative sources of energy, such as solar power. Development of locally made and efficient drip lines will eliminate huge import duties. Economic water productivity reduced with severity of deficit irrigation. It ranged from 1.57 US$ ha^{-1} mm^{-1} for T_{1101} to 2.42 US$ ha^{-1} mm^{-1} for T_{1111}.

The average seed yields of the most productive conservation practice during the rainy season, that was Mulch plus Soil bund (MLBD), was 19.2% higher than that of the full irrigation during the dry season. This shows that the average seasonal yield of 1 in 2 of the water conservation practices in the rainy season was higher than the yield under full irrigation in the dry season. Average seed yield of all the cultivation practices during the rainy season was 21.3% higher than that of the irrigated conditions. Similarly, the average water productivity of all the practices during the rainfed conditions was 14.3% higher than under full irrigation during the dry season and 29.1% higher than the average water productivity of all the irrigation scenarios during the dry seasons. In addition, the cost of production under rainfed conditions was significantly lower than under irrigated conditions. Economic water productivity under rainfed conditions was higher than that of dry season farming. For instance, seasonal output for Soil bund plus Mulch in the rainy season was 3.90 US$ ha^{-1} mm^{-1} of evapotranspired water while for full irrigation, that had the maximum seed yield in the dry season, it was 2.42 US$ ha^{-1} mm^{-1}. The revenue output for the conventional method was 2.27 US$ ha^{-1} mm^{-1} of ET_a under rainfed conditions and skipping irrigation during seed filling was 1.57 US$ ha^{-1} mm^{-1}. These facts and figures show that the productivity in terms of physical mass and economic value (revenue per unit of water evapotranspired) for the crop was higher under rainfed conditions than for the dry seasons. Higher water productivity under rainfed scenarios in this study is agreement with the finding that there is larger opportunity for improving water productivity under rainfed systems than with irrigated agriculture (Rockstrom, 2003; Oweis et al., 1998).

There are numerous benefits of increasing water productivity through higher yields and improved water use. It enhances crop production, generates and stabilizes income of the farmers, provides job opportunities, reduces price of commodities and reduces cost of production (Cai et al., 2011). The use of the water conservation practices in the current study will provide parts of these benefits. Cost of production under water conservation depends on type of materials used and the level of sophistication desired.

In addition, this study provides information on the economic productivity of water, which will serve as yardstick for planning water management for agriculture in the study area and within the Ogun-Osun River Basin.

Eighty per cent of the world's agricultural land area is rainfed and generates 58% of the world's staple foods (Stockholm International Water Institute (SIWI), 2001). In Sub-Saharan Africa (SSA) to which Nigeria belongs, more than 95% of the farmed land is rainfed. Land and water suitable for agriculture are limited on a global scale. There are some projections in the developing and developed countries for expansion of arable land, but by only up to 8%. Expansion of the arable land for agriculture will remain a vision in many developing countries in Sub-Saharan Africa, because of the huge investment involved, which is unsustainable under the current economic policies. Thus the focus for efforts to expand food production in the developing countries would have to be on raising crop yields on existing arable lands and improving production efficiencies, outcomes that can only be achieved by using improved cultivars together with improved agronomic practices (Cassman et al., 2003; Fereres, 2011). Agronomic practices especially under rainfed conditions need to be designed to improve the water productivity. Improving water productivity requires vapour shift (transfer) whereby soil physical conditions, soil fertility, crop varieties and agronomy are combined and managed to shift the evaporative loss into useful transpiration by plants (Evett and Tolk, 2009). This is a particular opportunity in water limiting regions of the world (Rockstrom et al., 2007).

Despite large dependence on rainfed agriculture for crop production, irrigated agriculture still offers promising benefits. About 18% of world's arable land is under irrigation and is responsible for about 40% of the crop output (International Commission on Irrigation and Drainage (ICID), 2006). Effort is being made where possible to shift to irrigated agriculture. The reason is that irrigated agriculture offers immunity against erratic or inadequate rainfall (Murty, 2008). In the study area, rainy season lasts for about 7 to 8 months but in the recent times, there have been fluctuations. Therefore, total reliance on rainfall without adequate preparation for either supplemental or total irrigation when needed will make the area to be more vulnerable to effects of drought. Under drip irrigation, water productivity is relatively high, water and fertilizer can be re-used. However, difficulties in adopting this system apart from the higher initial cost are the durability of the components and blockage of the outlets and poor functioning of the system. Consequently, if there is no modern technology to ensure effective operation and adequate maintenance, productivity will be very low and the huge investment made on drip system may not be justified (Murty, 2008). In the current study, a major challenge to the large-scale use of drip system in the cultivation of the crop was the cost (Table 8.9). Every effort made in minimizing the initial cost of production may increase the economic benefits. These measures include production of the drip lines with cheaper and durable materials and more importantly, local production. Others are the use of renewable energy such as solar power in pumping water either from the underground or surface sources. A major challenge to the conjunctive use of water for crop production in Nigeria and other developing economies is the large investment that it requires. Increasing water productivity is good under both rainfed and irrigated agriculture. It should not be pursued in isolation but in the context of achieving optimum balance between crop productivity and water use in the time of water scarcity in the study area.

Nitrogen is a unique element as we have it in abundance in our environment but its availability in forms suitable for plant use has made it an expensive input in crop production systems. Nitrogen input and its efficient use by plants is integral both to increasing crop production as well as to addressing issues of sustainability (Riar and

Coventry, 2013). In addition, the concentration of carbon dioxide in the atmosphere has been found to affect evapotranspiration and water use pattern of crops (Kirkham, 2011). Due to climate change and industrialization, concentration of carbon dioxide in the air is increasing every day and this may have significant effects on yield of Soybeans in the study area. Therefore, it is recommended that research be carried out on the effects of Nitrogen on yield components and threshold at which elevated carbon dioxide will affect evapotranspiration, yield and water productivity of the crop in the study area.

11. References

Abedinpour, M., Sarangi, A., Rajput, T.B.S., Singh, M., Pathak, H., Ahmad, T., 2012. Performance evaluation of AquaCrop model for maize crop in a semi-arid environment. *Agricultural Water Management*, 110(0): 55-66. DOI: http://dx.doi. org/10.1016/j.agwat.2012.04.001.

Adeboye, O.B., 2005. Flood Characteristics and Potential Reservoir Capacity of River Osun at Apoje Gauging Station. M. Eng Thesis, Unpublished, Federal University of Technology, Akure, Nigeria.

Adeboye, O.B., Alatise, O.M., 2008. Surface Water Potential of the River-Osun at Apoje Sub-basin Nigeria. *Soil Water Res*, 3(2): 74-79.

Adekalu, K.O., Osunbitan, J.A., Ojo, O.E., 2002. Water sources and demand in South Western Nigeria: implications for water development planners and scientists. *Technovation*, 12(12): 799-805. DOI: 10.1016/S0166-4972(01)00056-6.

Adekalu, K.O., 2004. Adapting Crop-Yield Models to Irrigation Scheduling in Nigeria, *Food Reviews International*, 20(4): 309-328. DOI: 10.1126/science.248.4954.477.

Adekalu, K.O., Okunade, D.A., 2008. Evaluation of crop yield models for cowpea in Nigeria, *Irrigation Science*, 26: 385-393. DOI: 10.1007/s00271-008-0103-6.

Adeniyan, O.N. and Ayoola, O.T., 2006. Growth and yield performance of some improved Soybean varieties as influenced by intercropping with maize and cassava in two contrasting locations in Southwest Nigeria, *African Journal of Biotechnology*, 5(20): 1886-1889.

Aduloju, M.O., Mahamood, J., Abayomi, Y.A., 2009. Evaluation of Soybean *(Glycine max (L) Merrill)* genotypes for adaptability to a southern Guinea savanna environment with and without P fertilizer application in north central Nigeria, *African Journal of Agricultural Research*, 4(6): 556-563.

Agudelo, O., Green, D.E., Shibles, R., 1986. Methods of Estimating Developmental Periods in Soybean. *Crop Sci.*, 26(6): 1226-1230. DOI: 10.2135/cropsci 1986.00 11183X002600060031x.

Akande, S.R, Owolade, O.F. and Ayanwole J.A., 2007. Field evaluation of Soybean varieties at Ilorin in the southern guinea savannah ecology of Nigeria, *African Journal of Agricultural Research*, 2(8): 356-359.

Akindele, S.T., Adebo, A., 2004. The Political Economy of River Basin and Rural Development Authourity in Nigeria: A Retrospective Case Study of Owena-River Basin and Rural Development Authourity (ORBRDA). *J. Hum. Ecol.*, 16(1): 55-62.

Alados, I., Foyo-Moreno, I., Alados-Arboledas, L., 1996. Photosynthetically active radiation: measurements and modelling. *Agricultural and Forest Meteorology*, 78(1-2): 121-131. DOI: http://dx.doi.org/10.1016/0168-1923(95)02245-7.

Alatise, M.O., Adeboye, O.B., 2005. Utilizing land and water resources at Apoje sub-basin (South-Western Nigeria) on Osun River for agricultural production, *Journal of Applied Science, Engineering and Technology*, 5(1 and 2): 75-79.

Alcamo, J., Doll, P., Kaspar, F., Siebert, S., 1997. Global Change and Global Scenarios of Water Use and Availability-An Application of Water GAP 1.0. University of Kassel, Centre for Environmental Systems Research, Kassel, Germany.

Allen, R.G., Pereira, L.S., Raes, D., Smith, M., 1998. Crop Evapotranpiration: Guidelines for computing crop water requirements, FAO Irrigation and Drainage Paper No. 56, Food and Agriculture Organization (FAO), Land and Water Development Division, Rome, Italy.

Ali, M.H., 2010. Fundamentals of Irrigation and On-Farm water management, Volume 1, *Springer Science*.

Alton, P.B., North, P.R., Los, S.O., 2007. The impact of diffuse sunlight on canopy light-use efficiency, gross photosynthetic product and net ecosystem exchange in three forest biomes" Global Change Biology, 13(4): 776-787.

Andriani, J., Andrade, F., Suero, E., Dardanelli, J., 1991. Water deficits during reproductive growth of soybeans. I. Their effects on dry matter accumulation, seed yield and its components. *Agronomie,* 11(9): 737-746.

Angus, J., Nix, H., Russell, J., Kruizinga, J., 1980. Water use, growth and yield of wheat in a subtropical environment. *Australian Journal of Agricultural Research,* 31(5): 873-886. DOI: http://dx.doi.org/10.1071/AR9800873.

Angus, J.F., van Herwaarden, A.F., 2001. Increasing Water Use and Water Use Efficiency in Dryland Wheat. *Agron. J.,* 93(2): 290-298. DOI:10.2134/agronj2001.932290x.

Ararso, G.S., 2005. Contribution of water management measures to food production in the Sub-Saharan Africa Region. MSc. thesis, UNESCO-IHE, Delft, the Netherlands.

Araya, A., Habtu, S., Hadgu, K.M., Kebede, A., Dejene, T., 2010. Test of AquaCrop model in simulating biomass and yield of water deficient and irrigated barley (*Hordeum vulgare*). *Agricultural Water Management,* 97(11): 1838-1846. DOI: doi.o rg/10.1016/j.agwat.2010.06.021.

Areola, O., Faniran, A., Akintola, O., 1985. The farmer-based small-farm schemes of the Ogun-Oshun River Basin Development Authority, Southwestern Nigeria. *Agricultural Systems,* 16(1): 7-21. DOI: http://dx.doi.org /10.1016/0308-521X(85)90006-X.

Arkebauer, T.J., Weiss, A., Sinclair, T.R., Blum, A., 1994. In defense of radiation use efficiency: a response to Demetriades-Shah et al. 1992. *Agricultural and Forest Meteorology,* 68(3-4): 221-227. DOI: http://dx.doi.org/10.1016/0168-1923(94)90038-8.

Ashley, D.A., Ethridge, W.J., 1978. Irrigation Effects on Vegetative and Reproductive Development of Three Soybean Cultivars. *Agron. J.,* 70(3): 467-471. DOI:10.2134/agronj1978.00021962007000030026x.

Atwell, B.J., Eamus, D., Bieleski, R.L., Farquhar, G., 1999. Plants in Action, Adaptation in nature, performance in action" Macmillan Education Australia, www.plants inaction.science.uq. edu.au (accessed on 23 August, 2013).

Bana, S., Prijono, S.A., Marno, S., 2013. The effects of Soil management on the Availability of Soil Moisture and Maize Production in Dryland, International *Journal of Agriculture and Forestry,* 3(3): 77-85.

Baird, J.R., Gallagher, J.N., Reid, J.B., 1987. Modelling the influence of flood irrigation on wheat and barley yields: A comparison of nine different models. *Adv. Irrig.* 4, 243-305.

Barber, S.A., 1978. Growth and Nutrient Uptake of Soybean Roots Under Field Conditions. *Agron. J.,*70(3): 457-461. DOI: 10.2134/agronj1978.00021962007000 030023x.

Belmans, C., Wesseling, J., Feddes, R., 1983. Simulation model of the water balance of a cropped soil: SWATRE. *Journal of Hydrology,* 63(3): 271-286.

Bennett, O.L., Ashley, D.A., Doss, B.D., 1966. Cotton Responses to Black Plastic Mulch and Irrigation . *Agron. J.,* 58(1): 57-60. DOI:10.2134/agronj1966.00 02196 2005800010019x.

Bhatia, V.S., Singh, P., Wani, S.P., Chauhan, G.S., Rao, A.V.R. K., Mishra, A.K., Srinivas, K., 2008. Analysis of potential yields and yield gaps of rainfed soybean in India using CROPGRO-Soybean model. *Agricultural and Forest Meteorology,* 148(8-9): 1252-1265. DOI: http://dx.doi.org/10.1016/j.agrformet.2 008.03.004.

Black, C., Ong, C., 2000. Utilisation of light and water in tropical agriculture. *Agricultural and Forest Meteorology,* 104(1): 25-47. DOI: http://dx.doi.org /10.10 16/S0168-1923(00)00145-3.

Blum, A., 1996. Crop responses to drought and the interpretation of adaptation. *Plant Growth Regulation,* 20(2): 135-148. DOI: 10.1007/BF00024010.

Boerma, H.R., Specht, J.E., 2004. *Soybeans: improvement, production and uses*: American Society of Agronomy.

Bonan, G., 2002. Ecological Climatology: Concepts and application, Cambridge University Press, 690 pp., ISBN:0521804760.

Bonhomme, R., 2000. Beware of comparing RUE values calculated from PAR vs solar radiation or absorbed vs intercepted radiation. *Field Crops Research,* 68(3): 247-252. DOI: http://dx.doi.org/10.1016/S0378-4290(00)00120-9.

Bos, M.G., Nugteren, J., 1974. On Irrigation Efficiencies, 1st edn. International Institute for Land Reclamation and Improvement: Wageningen, the Netherlands.

Bos, M.G., Nugteren, J., 1982. On Irrigation Efficiencies, 3rd edn. International Institute for Land Reclamation and Improvement: Wageningen, the Netherlands.

Brady, R.A., Stone, L.R., Nickell, C.D., Powers, W.L., 1974. Water conservation through proper timing of Soybeans irrigation. *J. Soil Water Cons.* 29: 266-268.

Brevedan, R.E., Egli, D.B., 2003. Short Periods of Water Stress during Seed Filling, Leaf Senescence, and Yield of Soybean *Crop Sci.,* 43(6): 2083-2088. DOI: 10.21 35/cropsci.2003.2083.

Briggs, I.J., Shantz, H.I., 1916. Daily transpiration during the normal growth period and its correlation with the weather. *Journal of Agric. Res.,*7: 155-212.

Brim, C.A., 1973. Quantitative genetics and breeding. In: ed. B.S. Caldwell, Soybeans: improvement, Production, and Uses. Agron. Mongr. 16, 1st ed. ASA, CSSA, and SSSA, Madison, WI, 155-186.

Bruinsma, J. (ed.), 2003. World Agriculture: Towards 2015/2030 An FAO Perspective. Earthscan and Food and Agriculture Organization, London and Rome, 444 pp

Burton, J.W., 1997. Soyabean (*Glycine max (L.)* Merr*.). Field Crops Research,* 53(1-3): 171-186. DOI: http://dx.doi.org/10.1016/S0378-4290(97)00030-0.

Cai X., Rosegrant M.W., 2003. World water productivity: current situation and future options. In: Kijne J.W., Barker R.M. (eds). Water productivity in Agriculture: Limits and opportunity for improvement International Water Management Institute (IWMI), Colombo, Sri Lanka: 163-178.

Cai, X., Molden, D., Mainuddin, M., Sharma, B., Ahmad, M.-u-D., Karimi, P., 2011. Producing more food with less water in a changing world: assessment of water productivity in 10 major river basins. *Water International,* 36(1): 42-62. DOI: 10.1080/02508060.2011.542403.

Candogan, B.N., Sincik, M., Buyukcangaz, H., Demirtas, C., Goksoy A.T., Yazgan, S., 2013. Yield, quality and crop water stress index relationships for deficit-irrigated Soybean *(Glycine max (L.) Merr.)* in sub-humid climatic conditions, *Agricultural Water Management,* 118(0): 113-121. DOI: dx.doi.org/10.1016/j.agwat.2012.11.021.

Campbell, G.S., Norman, J.M., 1998. An Introduction to Environmental Biophysics, 2nd ed. Springer-Verlag, New York, USA.

Carlson, J.B., 1973. Morphology. In: ed. B.E. Caldwell, Soybeans: Improvement, Production, and Uses. Agron. Mongr. 16, 1st ed. ASA, CSSA, and SSSA, Madison, WI: 17-95.

Caroline, S., 2002. Calculating a water poverty Index. *World Development*, 30(7): 1195-1210. DOI: 10.1016/S0305-750X(02)00035-9.

Cassman, K.G., Dobermann, A., Walters, D.T., Yang, H., 2003. Meeting cereal demand while protecting natural resources and improving environmental quality. *Annual Review of Environment and Resources,* 28(1): 315-358.

Ceotto, E., Di Candilo, M., Castelli, F., Badeck, F.-W., Rizza, F., Soave, C., Marletto, V., 2013. Comparing solar radiation interception and use efficiency for the energy crops giant reed (*Arundo donax L.*) and sweet sorghum (*Sorghum bicolor L. Moench*). *Field Crops Research,* 149(0): 159-166. DOI: http://dx.doi.org/10.1016/j.fcr.2013.05.002.

Chaudhary, M.R., Prihar, S.S., 1974. Root Development and Growth Response of Corn Following Mulching, Cultivation, or Interrow Compaction. *Agron. J.,* 66(3): 350-355. DOI: 10.2134/agronj1974.00021962006600030004x.

Commission on Sustainable Development (CSD), 2002. World Summit on Sustainable Development. 26 August - 4 September. Johannesburg, South Africa.

Confalone, A.E., Costa, L.C., Pereira, C.R., 1998. Growth and light capture in soybean under water stress, *Brazilian Journal of Agrometeorology*, 6(2): 165-169.

Confalone, A.E., Dujmovich, M.N., 1999. Influence of "deficit" water on the efficiency of solar radiation on soybean, *Brazilian Journal of Agrociencia*, 5(1): 195-198.

Constable, G.R., Hern, A.B., 1978. Agronomic and physiological response of Soybean and sorghum crops to water deficits. I. Growth, development and yield. *Aust. J. Plant Phyisol.,* 5: 159-167.

Cooper, P.J.M., Keatinge, J.D.H., Hughes, G., 1983. Crop evapotranspiration-a technique for calculation of its components by field measurements. *Field Crops Research,* 7(0): 299-312. DOI: http://dx.doi.org/10.1016/0378-4290(83)90038-2.

Cooper, R.L., Waranyuwat, A., 1985. Effect of Three Genes (Pd, Rpsal, and ln) on Plant Height, Lodging, and Seed Yield in Indeterminate and Determinate Near-Isogenic Lines of Soybeans. *Crop Sci.,* 25(1): 90-92. DOI: 10.2135/cropsci19 8 5.0011183X002500010023x.

Cordova, J.R., Bras, R.L., 1981. Physically based probabilistic models of infiltration, soil moisture, and actual evapotranspiration. *Water Resources Research,* 17(1): 93-106. DOI: 10.1029/WR017i001p00093.

Das, H.P., 2003. Water Use Efficiency of Soybean and its Yield Response to Evapotran -spiration and Rainfall, *Jour. Agric. Physics*, 3(1 & 2): 35-39.

Daughtry, C.S.T., Gallo, K.P., Goward, S.N., Prince, S.D., Kustas, W.P., 1992. Spectral estimates of absorbed radiation and phytomass production in corn and soybean canopies. *Remote Sensing of Environment,* 39(2): 141-152. DOI: http://dx.doi.org/10.1016/0034-4257(92)90132-4.

David, M.M., Coelho, D., Barrote, I., Correia, M.J., 1998. Leaf age effects on photosynthetic activity and sugar accumulation in droughted and rewatered *Lupinus albus* plants. *Functional Plant Biology,* 25(3): 299-306. DOI: http://dx.doi.org/10.1071/PP97142.

Decoteau, D.R., 2005. Principles of plant science. *Environmental Factors and Technology in Growing Plants*. Pearson and Prentice Hall: 49-66.

De Costa, W.A.J.M., Shanmugathasan, K.N., 2002. Physiology of yield determination of soybean (*Glycine max (L.) Merr.*) under different irrigation regimes in the sub-humid zone of Sri Lanka. *Field Crops Research*, 75(1): 23-35. DOI: http://dx.doi.org/10.1016/S0378-4290(02)00003-5.

de Fraiture, C., 2007. Integrated water and food analysis at the global and basin level. An application of WATERSIM, *Water Resources Management*, 21(1): 185-198. DOI: 10.10 07 /s11269-006-9048-9.

Demetriades-Shah, T.H., Fuchs, M., Kanemasu, E.T., Flitcroft, I., 1992. A note of caution concerning the relationship between cumulated intercepted solar radiation and crop growth. *Agricultural and Forest Meteorology*, 58(3-4): 193-207. DOI: http://dx.doi.org/10.1016/0168-1923(92)90061-8.

de Souza, P.I., Egli, D.B., Bruening, W.P., 1997. Water Stress during Seed Filling and Leaf Senescence in Soybean. *Agron. J.*, 89(5): 807-812. DOI: 10.2134/ agronj1997.00021962008900050015x.

de Souza, P.J.O.P., Ribeiro, A., Da Rocha, E.D.J.P., Farias, J.R.B., Loureiro, R.S., Bispo, C.C., Sampaio, L., 2009. Solar radiation use efficiency by soybean under field conditions in the Amazon region", Pesq. agropec. bras., Brasília, 44(10): 1211-1218.

De Wit, C. T., 1959. Potential photosynthesis of crop surfaces. *Netherlands* J. *Agric. Sci. 7*: 141-149.

De Wit, C.T., 1968. Simulation of Assimilation Respiration and Transpiration of Crops. John Wiley & Sons. New York, USA.

Dogan, E., Kirnak, H., Copur, O., 2007a. Deficit irrigations during soybean reproductive stages and CROPGRO-soybean simulations under semi-arid climatic conditions. *Field Crops Research*, 103(2): 154-159. DOI: http://dx.doi.org/10.1016/j.fcr.2007.05.009.

Dogan, E., Kirnak, H., Copur, O., 2007b. Effect of seasonal water stress on soybean and site specific evaluation of CROPGRO-Soybean model under semi-arid climatic conditions. *Agricultural Water Management*, 90(1-2): 56-62. DOI: http:// dx.doi.org/10.1016/j.agwat.2007.02.003.

Dornbos, D.L., Mullen, R.E., Shibles, R.E., 1989. Drought Stress Effects During Seed Fill on Soybean Seed Germination and Vigor. *Crop Sci.*, 29(2): 476-480. DOI: 10.2135/cropsci1989.0011183X002900020047x.

Doorenbos, J., Kassam, A.H., 1979. Yield response to water: *FAO* Irrigation and Drainage Paper No. 33, Rome, Italy.

Dugje, I.Y., Omoigui, L.O., Ekeleme F., Bandyopadhyay, R., Lava Kumar, P., Kamara, A.Y., 2009. Farmers' Guide to Soybean Production in Northern Nigeria. International Institute of Tropical Agriculture, Ibadan, Nigeria.

Eck, H.V., Mathers, A.C., Musick, J.T., 1987. Plant water stress at various growth stages and growth and yield of soybeans. *Field Crops Research*, 17(1): 1-16. DOI: http://dx.doi.org/10.1016/0378-4290(87)90077-3.

Egli, D.B., Orf, J.H., Pfeiffer, T.W., 1984. Genotypic Variation for Duration of Seedfill in Soybean1. *Crop Sci.*, 24(3): 587-592. DOI: 10.2135/cropsci1984 .001 118 3X 002400030037x.

Egli, D.B., 1994. Cultivar Maturity and Reproductive Growth Duration in Soybean. *Journal of Agronomy and Crop Science*, 173(3-4): 249-254. DOI: 10.1111/j.1439-037X.1994.tb00561.x.

Egli, D.B., 2004. Seed- Fill Duration and Yield of Grain Crops, *Advances in Agronomy* (Vol. Volume 83, 243-279): Academic Press.

Egli, D., 2005. Flowering, pod set and reproductive success in soya bean. *Journal of Agronomy and Crop Science*, 191(4): 283-291.

Elmore, R.W., Eisenhauer, D.E., Specht, J.E., Williams, J.H., 1988. Soybean yield and yield component response to limited capacity sprinkler irrigation systems. *J. Prod Agric.* 1: 196-201.

English, M.J., Musick, J.T., Murty, V.V., 1990. Deficit irrigation. In: G.J. Hoffman, T.A. Towell, K.H. Solomon, eds. *Management of farm irrigation systems*, ASAE. St. Joseph, Michigan, USA.

Evett, S.R., Tolk, J.A., 2009. Introduction: Can Water Use Efficiency Be Modeled Well Enough to Impact Crop Management? *Agron. J.,* 101(3): 423-425. DOI: 10.2134/ag ronj2009.0038xs.

Falkenmark, M., Rockstrom, J., 2004. Balancing Water for Humans and Nature: The New Approach in Ecohydrology. Earthscan, London, United Kingdom.

Farahani, H.J., Izzi, G., Oweis, T.Y., 2009. Parameterization and Evaluation of the AquaCrop Model for Full and Deficit Irrigated Cotton, *Agron. J.,* 101(3): 469-476. DOI:10.2134/agronj2008.0182s.

Federal Ministry of Environment, Nigeria (FMEN), 2010. Flood Research Documents in River Basins of Nigeria. Available on http://nigeriafews. net/flood research/ maps/riverbasin.php (accessed on 25 August, 2010).

Fehr, W.E., Caviness, C.E., 1977. Stages of Soybean Development. Cooperative Extension Service and Agriculture and Home Economics Experiment Station. Special Report 80. Iowa State University, Ames, IA, USA.

Fereres, E, Soriano, M.A., 2007. Deficit irrigation for reducing agricultural water use. *Journal of Experimental Botany,* 58(2): 147-159. DOI:10.1093/jxb/erl165.

Fereres, E., 2011. Optimizing water productivity in food production. In: Garrido, A., Ingram, H. (eds). Water for food in a changing world, Routledge, 14-32.

Fischer, G., Tubiello, F. N., van Velthuizen, H., Wiberg, D.A., 2007. Climate change impacts on irrigation water requirements: Effects of mitigation, 1990-2080. *Technological Forecasting and Social Change,* 74(7): 1083-1107. DOI: http://dx.doi.org/10.1016/j.techfore.2006.05.021.

Fletcher, A.L., Sinclair, T.R., Allen, L.H., 2007. Transpiration responses to vapour pressure deficit in well watered "slow-wilting" and commercial soybean. *Environmental and Experimental Botany,* 61: 145-151.

Flenet, F., Kiniry, J.R., Board, J.E., Westgate, M.E., Reicosky, D.E., 1996. Row Spacing Effects on Light Extinction Coefficients of Corn, Sorghum, Soybean, and Sunflower, *Agr. Journal,* 88: 185-190.

Food and Agriculture Organisation of the United Nations (FAO), 1999. Statistical Database (Online). Retrieved 2 June, 2010, from FAO Website: http://apps. Fao.org.

Food and Agriculture Organization of the United Nations (FAO), 2005. Irrigation in Africa in Figures, AquaStat Survey, Land and Water Management Division, FAO Report 29, Rome, Italy.

Food and Agriculture Organization of the United Nations (FAO), 2008. Hot Issues: water & poverty, an issue of life & livelihoods FAO web link: http://www.fao.org /nr/ water /issues/scarcity.html (accessed on 25 June, 2014).

Food and Agriculture Organisation of the United Nations (FAO), 2009. Statistics on Food Security: Aquastat Database: http://www. Fao.org/faostat/food security/in dexen.htm, Food and Agriculture Organization of the United Nations (FAO) (accessed on 10 April, 2010).

Food and Agriculture Organization of the United Nations (FAO), 2010a. FAOSTAT statistical database (Online). Available at http://apps.fao.org/ (accessed on 7 June, 2010).

Food and Agriculture Organization of the United Nations (FAO), 2010b. Water at a glance: The relationship between water, agriculture, food security and poverty. Available at http://www.fao.org/nr/water/infores.html (accessed on 27 July, 2010).

Food and Agriculture Organization of The United Nations (FAO), 2010c. AQUASTAT Country database http://www.fao.org/nr/water/aquastat/countries/Nigeria /index.stm (accessed on 2 June, 2010).

Food and Agriculture Organization of the United Nations (FAO), 2013. Online statistical data base of Food and Agriculture Organization of the United Nations, (accessed on 26 September, 2013).

Food and Agricultural Organization of the United Nations (FAO), 2014. FAO food price index. http://www.fao.org /worldfoodsitu ation/foodpricesindex/en/. (accessed on 20 June, 2014).

Framji, K., Garg, B., Luthra, C., 1982. Irrigation and drainage in the World.-A global review, 3rd ed. International Commission on Irrigation and Drainage (ICID), New Delhi, India.

French, R., Schultz, J., 1984a. Water use efficiency of wheat in a Mediterranean-type environment. I. The relation between yield, water use and climate. *Australian Journal of Agricultural Research, 35*(6): 743-764. DOI: http://dx.doi.org/ 10.10 71/AR9840743.

French, R., Schultz, J., 1984b. Water use efficiency of wheat in a Mediterranean-type environment. II. some limitations to efficiency. *Australian Journal of Agricultural Research,* 35(6): 765-775. DOI: http://dx .doi.o rg/10.1071/AR9840 765.

Gallagher, J.N., Biscoe, P.V., 1978. Radiation absorption, growth and yield of cereals. *J. Agric.Sci. Cambridge,* 91(1): 47-60.

Gbikpi, P.J., Crookson, R.K., 1981. A Whole-plant Indicator of Soybean Physiological Maturity. *Crop Sci.,* 21(3): 469-472. DOI: 10.2135/cropsci1981.001 1183X002100030030x.

Geerts, S., Raes, D., 2009. Deficit irrigation as an on-farm strategy to maximize crop water productivity in dry areas. *Agricultural Water Management,* 96(9): 1275-1284. DOI: 10.1016/j.agwat.2009.04.009.

Gercek, S., Boydak, E., Okant, M., Dikilitas, M., 2009. Water pillow irrigation compared to furrow irrigation for Soybean production in a semi-arid area, *Agricultural Water Management,* 87-92. DOI: doi.org/10.1016/j.agwat.2008.06.006.

Giller, K.E., Dashiel, K.E., 2007. *Glyxine Max (L.) Merr.* In: van der Vossen, H.A.M., Mkamilo, G.S. (eds). Plant Resources of Tropical Africa 14. Vegetable oils. PROTA Foundation, Wageningen, the Netherlands/Backhuys Publishers, Leiden, Netherlands/CTA: 78-84. Wageningen, the Netherlands.

Gosse, G., Varlet-Grancher, C., Bonhomme, R., Chartier, M., Allirand, J.-M., Lemaire, G., 1986. Maximum dry matter production and solar radiation intercepted by a canopy. *Agronomie,* 6: 47-56.

Hammer, G.L., Dong, Z., McLean, G., Doherty, A., Messina, C., Schussler, J., Zinselmeier, C., Paskiewicz, S., Cooper, M., 2009. Can Changes in Canopy and/or Root System Architecture Explain Historical Maize Yield Trends in the U.S. Corn Belt? *Crop Sci.,* 49(1): 299-312. DOI: 10.2135/cropsci2008.03.0152.

Hanks, R.J., Gardner, H.R., Florian, R.L., 1969. Plant Growth-Evapotranspiration Relations for Several Crops in the Central Great Plains. *Agron. J.,* 61(1): 30-34. DOI: 10.2134/agronj1969.00021962006100010010x.

Hanks, R.J., 1974. Model for Predicting Plant Yield as Influenced by Water Use. *Agron. J.*, 66(5): 660-665. DOI: 10.2134/agronj1974.00021962006600050017x.

Hanks, R.J., Stewart, J.I., Riley, J.P., 1976. Four state comparisons of models used for irrigation management. Proc. ASCE. *Irrig*. Drainage Div: 283-294.

Hanks, R.J., Hill, R.W., 1980. Modelling Crop Response to Irrigation in Relation to Soils, Climate and Salinity. Pergamon, Oxford, United Kingdom.

Heath, M.C., Hebblethwaite, P.D., 1985. Solar radiation interception by leafless, semi-leafless and leafed peas (*Pisum sativum*) under contrasting field conditions. Ann. *App. Biol.* 107: 309-318.

Heatherly, L.G., 1983. Response of Soybean Cultivars to Irrigation of a Clay Soil. *Agron. J.,* 75(6): 859-864. DOI: 10.2134/agronj1983.00021962007500060004x.

Heatherly, L.G., 1993. Drought Stress and Irrigation Effects on Germination of Harvested Soybean Seed. *Crop Sci.,* 33(4): 777-781. DOI: 10.2135/cropsci1993. 0011183X003300040029x.

Henggeler, J.C., Enciso, J.M., Multer, W.L., Unruh, B.L., 2002. Deficit Subsurface Drip Irrigation of Cotton in: FAO 2002. Deficit Irrigation Practices, Water Reports 22: 29-38. Rome, Italy.

Hill, R.W, Hanks R.J., Wright J.L., 1982. Crop yield models adapted to irrigation schedules programs (CRPSM). Research Report 100; Utah: Utah State University, Logan, USA.

Hill, H.J., West, S.H., Hinson, K., 1986. Soybean Seed Size Influences Expression of the Impermeable Seed-Coat Trait. *Crop Sci.,* 26(3): 634-637. DOI: 10.2135/cro psci1986.0011183X002600030044x.

Hillel, D., Talpaz, H., Van Keulen, H., 1976. A macroscopic scale model of water up take by a non-uniform root system and of water and salt movement in the soil profile, *Soil Science* 121: 242-245.

Howell, T.A., Meek, D.W., Hatfield, J.L., 1983. Relationship of photosynthetically active radiation to shortwave radiation in the San Joaquin Valley. *Agricultural Meteorology,* 28(2): 157-175. DOI: 10.1016/0002-1571(83) 90005-5.

Howell, T.A., Cuenca, R.H., Solomon, K.H., 1990. Crop yield response. In: Hoffman, G.J., Howell, T.A., Solomon, K.H. (eds), Management of Farm Irrigation Systems. Am. Soc. Agric. Eng. (ASAE): 93-122. St. Joseph, MI, USA.

Howell, T.A., 2001. Enhancing Water Use Efficiency in Irrigated Agriculture. *Agron. J.,* 93(2): 281-289. DOI: 10.2134/agronj2001.932281x.

Hsiao, T.C., Heng, L., Steduto, P., Rojas-Lara, B., Raes, D., Fereres, E., 2009. AquaCrop-The FAO Crop Model to Simulate Yield Response to Water: III. Parameterization and Testing for Maize. *Agron. J.,* 101(3): 448-459. DOI: 10.21 34 /agronj2008.0218s.

Huck, M.G., Peterson, C.M., Hoogenboom, G., Bush, C.D., 1986. Distribution of dry mater between shoots and roots of irrigated and non-irrigated Soybeans. *Agron. J.* 78: 807-813.

Huck, M.G., Peterson, C. M., Hoogenboom, G., Busch, C.D., 1986. Distribution of Dry Matter Between Shoots and Roots of Irrigated and Non-irrigated Determinate Soybeans. *Agron.   J.,   78(5):   807-813.   DOI: 10.2134/agronj1986.0 0021962007800050013x.

Hughes, G., Keatinge, J.D.H., Scott, Cooper, J.B.M., Dee, N.F., 1987. Solar radiation interception and utilization by chickpea (*Cicer arietinum* L.) crops in northern Syria. *J. Agric. Sci.* Camb. 108: 419-424.

Hussein, F., Janat, M., Yakoub, A., 2011. Simulating cotton yield response to deficit irrigation with the FAO AquaCrop model. *Spanish Journal of Agricultural Research, 9*(4): 1319-1330.

Igbadun, E.H., Mahoo, F.H., Tarimo, A.K., Salim, B.A., 2006. Crop water productivity of an irrigated maize crop in Mkoji sub-catchment of the Great Ruaha River basin, Tanzania, *Agricultural Water Management*: 141-150. DOI: doi.org/10.10 16/j. ag wat.2006.04.003.

Igbadun, H.E., Tarimo, A.K.P.R, Salim, B.A., Mahoo, H.F., 2007. Evaluation of selected crop water production functions for an irrigated maize crop. *Agricultural Water Management*, 94: 1-10. DOI: 10.1016/j.agwat.2007.07.006.

International Commission on Irrigation and Drainage (ICID), 2006. Water for Food and the Environment-Main Issues and Recent Tendencies, Paper presented at the 4th World Water Forum, Mexico, 2006, 31-40.

International Commission on Irrigation and Drainage (ICID), 2008. Revised Draft Topics Scoping Paper Topic 2.3, Water and food for ending poverty and hunger. Prepared by Henri Tardieu and Bart Schultz. 5 October, New Delhi, India.

International Water Management Institute (IWMI), 2006. Beyond more crop per drop. Background paper prepared for 4th World Wide Water Forum, Mexico 2006. International Water Management Institute. Online at www.iwmi.Cgiar.org /WWF4/html/action_2.htm.

Irz, X., Roe, T., 2000. Can the world feed itself? Some insights from growth theory, Agrekon *Agrekon: Agricultural Economics Research, Policy and Practice in Southern Africa*, 2078-0400, 39(4): 513-528. DOI: 10.1080/030318 53.2000.9 52366.

Israelsen, O.W., 1950. Irrigation Principles and Practices, John Wiley and Sons, Inc., New York; USA.

Javaheri, F., Baudoin, J.P., 2001. Soya bean. In: Raemaekers, R.H. (ed.). Crop production in tropical Africa. DGIC (Directorate General for International Co-operation), Ministry of Foreign Affairs, External Trade and International Co-operation: 809-828. Brussels, Belgium.

Jensen, M.E., 1967. Evaluating irrigation efficiency. *Journal of Irrigation and Drainage* Division of American Society of Civil Engineers 93(IR1): 83-98.

Jensen, M.E., 1968. Water consumption by agricultural plants. In: Water Deficits and Plant Growth. Vol. II. Academic Press. New York, USA.

Jeuffroy, M.-H., Ney, B., 1997. Crop physiology and productivity. *Field Crops Research*, 53(1-3): 3-16. DOI: http://dx.doi.org/10.1016/S0378-4290(97)00019-1.

Jones, H.G., 1992. Plants and Microclimate: A Quantitative Approach to Environmental Plant Physiology, 2nd ed. Cambridge University Press, Cambridge, United Kingdom.

Jury, W.A., Vaux Jr, H.J., 2007. The Emerging Global Water Crisis: Managing Scarcity and Conflict Between Water Users. In Donald, L.S. (ed.), *Advances in Agronomy*, 95, 1-76): Academic Press.

Kadigi, R.M.J., Kashaigili, J.J., Mdoe, N.S., 2004. The economics of irrigated paddy in Usangu Basin in Tanzania: water utilization, productivity, income and livelihood implications. *Physics and Chemistry of the Earth, Parts A/B/C*, 29(15-18): 1091-1100. DOI: http://dx.doi.org/10.1016/j.pce.2004.08.010.

Karam, F., Masaad, R., Sfeir, T., Mounzer, O., Rouphael, Y., 2005. Evapotranspiration and seed yield of field grown Soybean under deficit irrigation conditions. *Agricultural Water Management*, 75(3): 226-244, DOI: 10.1016/jagwat.2004.12.0 15.

Keller, J., Bliesner, R.D., 1990. Sprinkler and trickle irrigation, Van Noststrand Reinhold, New York, USA.

Khan, A.R., 1996. Influence of tillage on soil aeration. *Journal of Agronomy and Crop Science,* 177(4): 253-259. DOI: 10.1111/j.1439-037X.1996.tb00243.x.

Khurshid, K., Iqbal, M.S., M.S., Arif and Nawaz, A., 2006. Effect of tillage and mulch on soil physical properties and growth of maize. *Int. J. Agric. Biol.,* 8(5): 593-596.

Kijne, J.W., Barker, R., Molden, D., 2003. Improving water productivity in agriculture: Editor's Overview. In: Kijne, J.W., Barker, R., Molden, D. (eds), *Water productivity in Agriculture: Limits and Opportunities for Improvement.* A Comprehensive Assessment of Water Management in Agriculture: xi-xix. Earthscan. London, United Kingdom and International Water Management Institute (IWMI). Colombo, Sri Lanka.

Kiniry, J.R., Jones, C.A., O'Toole, J.C., Blanchet, R., Cabelguenne, M., Spanel, D.A., 1989. Radiation-use efficiency in biomass accumulation prior to grain-filling for five grain-crop species. *Field Crops Research,* 20(1): 51-64. DOI: http://dx.doi.org/10.1016/0378-4290(89)90023-3.

Kiniry, J.R., Williams, J.R., Gassman, P.W., Debaeke, P., 1992. A general, process-oriented model for two competing plant species. Trans. ASAE 35: 801-810.

Kiniry, J.R., Tischler, C. R., van Esbroeck, G.A., 1999. Radiation use efficiency and leaf CO_2 exchange for diverse C4 grasses. *Biomass and Bioenergy,* 17(2): 95-112. DOI: http://dx.doi.org/10.1016/S0961-9534(99)00036-7.

Kirda, C, Moutonnet, P., Hera, C., Nielsen, D.R., 1999. Preface. In: Kirda, C., Montonnet, P., Hera, C., Nielsen, D.R. (eds). Crop yield response to deficit irrigation (vii-viii). Boston, London: Kluwer Academics Publishers. Dordrecht, the Netherlands.

Kirda, C., 2002. Deficit irrigation scheduling based on plant growth stages showing water stress tolerance. In: *Deficit irrigation practice,* Water Report No 22: 3-10. Food and Agriculture Organisation of the United Nations (FAO). Rome, Italy.

Kirkham, M.B., 2004. Principles of Soil and Plant Water Relations, Springer. Heidelberg, Germany.

Kirkham, M.B., 2011. Elevated carbon dioxide: impacts on soil and plant water relations: CRC Press: 197-223.

Klocke, N.L, Eisenhauer, D.E., Specht, J.E., Elmore, R.W., Hergert, G.W., 1989. Irrigation of Soybeans by growth stages in Nebraska. Trans ASAE 5: 361-366.

Koegelenberg, F.H., 2003. Irrigation design manual. Chapter 12. ARC - Institute for Agricultural Engineering, South Africa.

Korte, L.L., Williams, J.H., Specht, J.E., Sorensen, R.C., 1983. Irrigation of Soybean Genotypes During Reproductive Ontogeny. I. Agronomic Responses1. *Crop Sci.,* 23(3): 521-527. DOI: 10.2135/cropsci1983.0011183 X002300030019x.

Kraus, H.T., 1998. Effects of mulch on soil moisture and growth of Desert Willow, *Hort'Technology,* 8(4): 588-590.

Krause, P., Boyle, D.P., Bäse, F., 2005. Comparison of different efficiency criteria for hydrological model assessment. *Adv. Geosci.,* 5, 89-97. DOI: 10.5194/adge o-5-89-2005.

Lamm, F.R., Abou Kheira, A.A., Trooien, T.P., 2010. Sunflower, Soybean and grain Sorghum crop production as affected by drip line depth, American Society of Agricultural and Biological Engineers ISSN 0883-8542, 26(5): 873-882.

Latiri-Souki, K., Nortcliff, S., Lawlor, D.W., 1998. Nitrogen fertilizer can increase dry matter, grain production and radiation and water use efficiencies for durum wheat under semi-arid conditions. *European Journal of Agronomy,* 9(1): 21-34. DOI: http://dx.doi.org/10.1016/S1161-0301(98)00022-7.

Leadley, P.W., Reynolds, J.F., Flagler, R., Heagle, A.S., 1990. Radiation utilization efficiency and the growth of soybeans exposed to ozone: a comparative analysis. *Agricultural and Forest Meteorology*, 51(3-4): 293-308. DOI: http://dx.doi.org/ 10.1016/0168-1923(90)90114-L.

Lecoeur, J., Guilioni, L., 1998. Rate of leaf production in response to soil water deficits in field pea. *Field Crops Research*, 57(3): 319-328. DOI: http://dx.doi. org/10.1016 /S0378-4290(98)00076-8.

Lipton, M., Litchfield, J., Faures, J.M., 2003. The effects of irrigation on poverty: a framework for analysis. *Water Policy*, 5(5): 413-427.

Liu, T., Song, F., Liu, S., Zhu, X., 2012. Light interception and radiation use efficiency response to narrow-wide row planting patterns in maize, *Australian. Journal of Crop Science*, 6(3): 506-513.

Liu, S., Zhang, X.-Y., Yang, J., Drury, C.F., 2013. Effect of conservation and conventional tillage on soil water storage, water use efficiency and productivity of corn and soybean in Northeast China. *Acta Agriculturae Scandinavica, Section B-Soil & Plant Science*, 63(5): 383-394. DOI: 10.1080/09064710.2012.76 2803.

Loomis, R.S., Williams, W.A., 1963. Maximum crop productivity: an extimate. *Crop Sci.*, 3(1): 67-72. DOI: 10.2135/cropsci1963.0011183X000300010021x.

Loss, S.P., Siddique, K.H.M., Tennant, D., 1997. Adaptation of faba bean (Vicia faba L.) to dryland Mediterranean-type environments III. Water use and water-use efficiency. *Field Crops Research*, 54(2-3): 153-162. DOI: http: //dx .doi .org /10 .1016/S0378-4290(97)00042-7.

Lovelli, S., Perniola, M., Ferrara, A., Tommaso, D.T., 2007. Yield response factor to water (Ky) and water use efficiency of *Carthamus tinctorius L.* and *Solanum melongena L. Agricultural water management*, 73-80. DOI: dx.doi.org/10.1016/j.a gwat.2007.05.005.

Lu, X., Jin, M., van Genuchten, M.T., Wang, B., 2011. Groundwater Recharge at Five Representative Sites in the Hebei Plain, China. *Ground Water*, 49(2): 286-294. DOI: 10.1111/j.1745-6584.2009.00667.x.

Ma, G., Huang, J., Wu, W., Fan, J., Zou, J., Wu, S., 2013. Assimilation of MODIS-LAI into the WOFOST model for forecasting regional winter wheat yield. *Mathematical and Computer Modelling*, 58(3-4): 634-643. DOI: http://dx.doi.org/ 10.1016/j.mcm.2011.10.038.

Martin, C.K., Cassel, D.K., Kamprath, E.J., 1979. Irrigation and Tillage Effects on Soybean Yield in a Coastal Plain Soil. *Agron. J.*, 71(4): 592-594. DOI: 10.2134 /agronj1979.00021962007100040017x.

Martin, B., Kebede, H., Rilling, C., 1994. Photosynthetic Differences among Lycopersicon Species and *Triticum aestivum* Cultivars. *Crop Sci.*, 34(1): 113-11 8. DOI: 10.2135/cropsci1994.0011183X003400010020x.

Matson, P.A., Parton, W.J., Power, A.G., Swift, M.J., 1997. Agricultural Intensification and Ecosystem Properties. *Science*, 277(5325): 504-509. DOI: 10.1126/science.277.5325.504.

Mavi, H.S., Tupper, G., 2004. Agrocmeteorology: Principles and Applications of Climate Studies in Agriculture. Food Products Press. New York, USA.

McMaster, G.S., Wilhelm, W.W., 1997. Growing degree-days: one equation, two interpretations. *Agricultural and Forest Meteorology*, 87(4): 291-300. DOI: http: //dx.doi.org/10.1016/S0168-1923(97)00027-0.

Metz, G.L., Green, D.E., Shibles, R.M., 1985. Reproductive Duration and Date of Maturity in Populations of Three Wide Soybean Crosses. *Crop Sci.*, 25(1): 171-176. DOI: 10.2135/cropsci1985.0011183X002500010041x.

Michael, A.M., 2008. Irrigation, Theory and Practice, Vikas Publishing House PVT Ltd, New Delhi, India.

Minhas, B.S., Parkhand, K.S., Srinivasan, T.N., 1974. Towards the structure of a production function for wheat yields with dated input of irrigation water. *Water Resources Research*, 10(3): 383-386. DOI: 10.1029/WR010i003p00383.

Minhas, J.S., Bansal, K.C., 1991. Tuber yield in relation to water stress at stages of growth in potato (*Solanum Tuberosum* L.). *Journal of the Indian Potato Association* 18: 1-8.

Mkhabela, M.S., Bullock, P.R., 2012. Performance of the FAO AquaCrop model for wheat grain yield and soil moisture simulation in Western Canada. *Agricultural Water Management*, 110, 16-24. DOI: http://dxdoi.Org/10.1016/j.Agwat2012.0 3.009.

Mohammed, S.O., Farshad, A., Farifteh, J., 1996. Evaluating land degradation for assessment of land vulnerability to desert conditions in Sokoto area, Nigeria, *Land degradation and development*, 7(3): 205-215. DOI:10.1002/(SICI)1099-145X(199609).

Molden, D., Sakthivadivel R., 1999. Water accounting to assess use and productivity of water. International. *Journal of Water Resources Development*, 15 (1&2): 55-71. DOI: 10.1080/07900629948934.

Molden, D., Murray-Rust, H., Sakthivadivel, R., Makin, I., 2003. A Water Productivity Framework for Understanding and Action. In Kijne JW, Barker R and Molden, D. (eds). *Water productivity in agriculture: limits and opportunities for improvements*. A Comprehensive Assessment of Water Management in Agriculture: 1-18. Earthscan. London, United Kingdom and International Water Management Institute (IWMI). Colombo, Sri Lanka.

Molden, D., Frenken, K., Barker, R., de Fraiture, C., Mati, B., Svendsen, M., Sadoff, C., Finlayson, C.M., Attapatu, S., Giordano, M., Inocencio, A., Lannerstad, M., Manning, N., Molle, F., Smedema, B., Vallée, D., 2007. Trends in water and agricultural development. In: Molden, D. (ed.), *Water for food and water for life: A Comprehensive Assessment of Water Management in Agriculture: 1-18.* Earthscan. London, United Kingdom and International Water Management Institute (IWMI). Colombo, Sri Lanka.

Monteith, J.L., 1965. Light distribution and photosynthesis in field crops. Ann. Bot. 29: 17-37.

Monteith, J.L., 1975. Principles of Environmental Physics, Edward Arnold, London, United Kingdom.

Monteith, J.L., 1977. Climate and the efficiency of crop production in Britain, Philos. Trans. R. Soc., 281: 277-294. London, United Kingdom.

Monteith, J.L., 1981. Climatic variation and the growth of crops. *Quarterly Journal of the Royal Meteorological Society*, 107(454): 749-774. DOI: 10.1002/qj.497107 45402.

Monteith, J.L., 1994. Validity of the correlation between intercepted radiation and biomass. *Agricultural and Forest Meteorology*, 68(3-4): 213-220. DOI: http: //dx.doi.org /10.1016/0168-1923(94)90037-X.

Monteith, J.L., 1996. The Quest for Balance in Crop Modeling. *Agron. J.*, 88(5): 695-697. DOI: 10.2134/agronj1996.00021962008800050003x.

Monteith, J.L., Unsworth, M.H., 2013. Principles of Environmental Physics-Plants, Animals, and the Atmosphere, Elsevier: 49-78.

Momen, N.N., Carlson, R.E., Shaw, R.H., Arjmand, O., 1979. Moisture-Stress Effects on the Yield Components of Two Soybeans Cultivars. *Agron. J.,* 71(1): 86-90. DOI: 10.2134/agronj1979.00021962007100010022x.

Morgan, R.P.C., 1996. *Soil Erosion and Conservation.* 2nd edition Longman Group and John Willey & Sons: 40-46. New York, USA.

Moriasi, D., Arnold, J., van Liew, M., Bingner, R., Harmel, R., Veith, T., 2007. Model evaluation guidelines for systematic quantification of accuracy in watershed simulations. *Trans. ASABE,* 50(3): 885-900.

Muchow, R.C., 1985. An analysis of the effects of water deficits on grain legumes grown in a semi-arid tropical environment in terms of radiation interception and its efficiency of use. *Field Crops Research,* 11(0): 309-323. DOI: http://dx.doi.org /10.1016/0378-4290(85)90111-X.

Muchow, R.C., Sinclair, T.R., Bennett, J.M., Hammond, L.C., 1986. Response of Leaf Growth, Leaf Nitrogen, and Stomatal Conductance to Water Deficits during Vegetative Growth of Field-Grown Soybean1. *Crop Sci.,* 26(6): 1190-1195. DOI: 10.2135/cropsci1986.0011183X002600060024x.

Mukherjee, A., Kundu, M., Sarkar, S., 2010. Role of irrigation and mulch on yield, evapotranspiration rate and water use pattern of tomato (*Lycopersicon esculentum L.*). *Agricultural Water Management,* 98(1): 182-189. DOI: http://dx. doi.org/10.1016/j.agwat.2010.08.018.

Munn, D.A., 1992. Comparisons of shredded newspaper and wheat straw as crop mulches. *HortTechnology,* 2(3): 361-366.

Murty, V.V.N., 2008. Irrigated agriculture for food security. In: Aswathanarayana, U. (ed.). Food and water security. Taylor & Francis/Balkema: 77-87. Leiden, the Netherlands.

Musick, J.T., Jones, O.R., Stewart, B.A., Dusek, D.A., 1994. Water-Yield Relationships for Irrigated and Dryland Wheat in the U.S. Southern Plains. *Agron. J.,* 86(6): 980-986. DOI: 10.2134/agronj1994.00021962008600060010x.

Muthamilselvan, M., Manian, R., Kathirvel, K., 2006. In-situ Moisture Conservation Techniques in Dry farming-A Review, *Agric. Rev.,* 24(1): 67-72.

Nakeseko, K., Gotoh, K., 1983. Comparative studies on the dry matter production, plant type and productivity in Soybean, adzuki bean and kidney bean. VII: An analysis of the productivity among three crops on the basis of radiation absorption and its efficiency for dry matter accumulation. *Jpn. J. Crop Sci.* 52(1): 49-58.

Narayanan, S., Aiken, R.M., Vara Prasad, P.V., Xin, Z., Yu, J., 2013. Water and Radiation Use Efficiencies in Sorghum. *Agron. J.,* 105(3): 649-656. DOI: 10.21 34/agronj2012.0377.

Nash, J.E., Sutcliffe, J.V., 1970. River flow forecasting through conceptual models part I - A discussion of principles. *Journal of Hydrology,* 10(3): 282-290. DOI: ht tp://dx .doi.org/10.1016/0022-1694(70)90255-6.

National Millennium Development Goals (NMDG), 2004. Report on National Millennium Development Goals (www.undg.org) (accessed on 5 April, 2010).

National Population Commission of Nigeria (NPCN), 2010. Nigerian Population Facts and Figure, http://www.population.gov.ng/factsandfigures.htm (accessed on 13 July, 2010).

Nautiyal, P.C., Joshi, Y.C., Dayal, D., 2002. Response of groundnut to deficit irrigation during vegetative growth in: FAO 2002. Deficit Irrigation Practices, Water Reports 22: 39-47.

Neil, S.G., 1995. Water resources Planning, McGraw-Hill Inc, USA.

Neyshabouri, M.R., Hatfield, J.L., 1986. Soil water deficit effects on semi-determinate and indeterminate soybean growth and yield. *Field Crops Research,* 15(1): 73-84. DOI: http://dx.doi.org/10.1016/0378-4290(86)90102-4.

Nielsen, D.C., 1990. Scheduling irrigations for soybeans with the Crop Water Stress Index (CWSI). *Field Crops Research,* 23(2): 103-116. DOI: http://dx.doi.org/10.1016/0378-4290(90)90106-L.

Nigerian National Committee on Irrigation and Drainage (NINCID), 1999. Country Profile - Nigeria. Federal Ministry of Agriculture & Water Resource. Abuja, Nigeria.

O'Connell, M.G., O'Leary, G.J., Whitfield, D.M., Connor, D.J., 2004. Interception of photosynthetically active radiation and radiation-use efficiency of wheat, field pea and mustard in a semi-arid environment. *Field Crops Research,* 85(2-3): 111-124. DOI: http://dx.doi.org/10.1016/S0378-4290(03)00156-4.

Obalum, S.E., Igwe, C.A., Obi, M.E., Wakatsuki, T., 2011. Water use and grain yield response of rainfed soybean to tillage-mulch practices in southeaster Nigeria. Sci. *Agric* (Piracicaba, Braz.), 68(5): 554-561.

Ogun-Osun River Basin and Rural Development Authourity (OORBA), 1982. Feasibility Reports on Osun River Basin, Federal Government of Nigeria.

Othieno, C.O., 1980. Effects of Mulches on Soil Water Content and Water Status of Tea Plants in Kenya. *Experimental Agriculture,* 16:295-302. DOI: 10.1017/S00144797 00011054.

Oweis, T., Pala, M., Ryan, J., 1998. "Stabilizing rain-fed wheat yields with supplemental irrigation and nitrogen in a Mediterranean-type climate". *Agronomy Journal* 90, 672-681.

Oweis, T., Hachum, A., 2006. Water harvesting and supplemental irrigation for improved water productivity of dry farming systems in West Asia and North Africa. *Agricultural Water Management,* 80(1-3): 57-73. DOI: http://dx.doi.Org/1 0.1016/j.agwat.2005.07.004.

Oyekale, A.S., 2007. Determinants of Agricultural Land Expansion in Nigeria: An Application of Error Correction Modelling (ECM), *Central European Journal of Agriculture,* 8(3): 301-310.

Palmer, R.G., Kilen, T.C., 1987. Qualitative genetics and cytogenetics. In: ed. J.R. Wilcox, Soybeans: Improvement, Production, and Uses. Agron. Mongr. 16, 2nd ed. ASA, CSSA, and SSSA: 135-209. Madison, WI, USA.

Palosuo, T., Kersebaum, K.C., Angulo, C., Hlavinka, P., Moriondo, M., Olesen, J.E., Rotter, R., 2011. Simulation of winter wheat yield and its variability in different climates of Europe: A comparison of eight crop growth models. *European Journal of Agronomy,* 35(3): 103-114. DOI: http://dx.doi.org/10.1016/j.eja.2011.05.001.

Passioura, J.B., Angus, J.F., 2010. Improving productivity of crops in water-limited environments. In: Donald L.S. (ed.), Advances in Agronomy, 106: 37-75. Academic Press.

Payero, J.O., Melvin, S.R., Irmak, S., 2005. Response of Soybean to deficit irrigation in Semi-Arid Environment of West-Central Nebraska, Transactions of the ASAE, 48(6): 2189-2203.

Penman, H.L., 1962. Woburn irrigation, 1951-59 I. Purpose, design and weather. *The Journal of Agricultural Science,* 58(03): 343-348. DOI: 10.1017/S002185960 0013204.

Perry, C., 2007. Efficient Irrigation; Inefficient Communication; Flawed Recommendations, *Irrigation and Drainage,* 56: 367-378. DOI: 10.1002/ird.323.

Plenet, D., Mollier, A., Pellerin, S., 2000. Growth analysis of maize field crops under phosphorus deficiency. II. Radiation-use efficiency, biomass accumulation and yield components. *Plant and Soil,* 224(2): 259-272.

Postel, S., 1999. Pillars of Sand: Can the Irrigation Miracle Last? W.W. Norton, New York, USA.

Pradhan, S., Sehgal, V.K., Das, D.K., Jain, A.K., Bandyopadhyay, K.K., Singh, R., Sharma, P.K., 2014. Effect of weather on seed yield and radiation and water use efficiency of mustard cultivars in a semi-arid environment. *Agricultural Water Management,* 139(0): 43-52. DOI: http://dx.doi.org/10.1016/j.Agwat.2014.03.005.

Purcell, L.C., Ball, R.A., Reaper, J.D., Vories, E.D., 2002. Radiation Use Efficiency and Biomass Production in Soybean at Different Plant Population Densities. *Crop Sci.,* 42(1): 172-177. DOI: 10.2135/cropsci2002.172.

Purcell, L.C., Edwards, J.T., Brye, K.R., 2007. Soybean yield and biomass responses to cumulative transpiration: Questioning widely held beliefs. *Field Crops Research,* 101(1): 10-18. DOI: http://dx.doi.org /10.1016/j.fcr.2006.09.0 0 2.

Raes, D., Steduto, P., Hsiao, T.C, Fereres, E., 2009. AquaCrop: The FAO Crop Model to Simulate Yield Response to Water: II. Main Algorithms and Software Description, *Agronomy Journal,* 101(3): 438-447. DOI: 10.2134/agronj2008.0140s.

Raes, D., Steduto, P., Hsiao, T.C., Fereres, E., 2012. *AquaCrop* - Reference Manual. Available at: http://www.fao.org/nr/water/aquacrop.html.

Ray, L., Max, A.K., Joseph, L.H., 1988. Hydrology for engineers, International Edition, McGraw-Hill Inc, 1-3.

Reicosky, D.C., Deaton, D.E., 1979. Soybean Water Extraction, Leaf Water Potential, and Evapotranspiration During Drought1. *Agron. J.,* 71(1): 45-50. DOI: 10.2134/agronj1979.00021962007100010011x.

Ren, X., Jia, Z., Chen, X., 2008. Rainfall concentration for increasing corn production under semiarid climate. *Agricultural Water Management,* 95(12): 1293-1302. DOI: http://dx.doi.org/10.1016/j.agwat.2008.05.007.

Repetto, R., 1986. Skimming the water: rent-seeking and the performance of public irrigation systems. Research Report 4, World Resources Institute, Washington DC, USA.

Retta, A., Vanderlip, R.L., Higgins, R.A., Moshier, L.J., 1996. Application of SORKAM to Simulate Shattercane Growth Using Forage Sorghum. *Agron. J.,* 88(4): 596-601. DOI: 10.2134/agronj1996.00021962008800040017x.

Riar A., and Coventry, D., 2013. Nitrogen Use as a Component of Sustainable Crop Systems. In:Bhullar G.S and Bhullar, N.K. (eds). Agricultural Sustainability-Progress and Prospects in Crop Research, Academic Press: 63-76.

Rickard, D., Fitzgerald, P., 1969. The estimation and occurrence of agricultural drought. *Journal of Hydrology (*New Zealand*),* 8: 11-16.

Ritchie, J.T., 1972. Model for predicting evaporation from a row crop with incomplete cover. *Water Resources Research,* 8(5): 1204-1213. DOI: 10.1029/WR008i005 p 01204.

Ritchie, J.T, Otter, S., 1985. Description and Performance of CERES-Wheat. A User-Oriented Wheat Yield Model. U.S. Dept. of Agriculture, Temple, Texas, USA.

Ritchie, J.T., Godwin, D.C., Otter-Nacke, S., 1985. CERES-Wheat: A Simulation Model of Wheat Growth and Development. Texas A. & M Univ. press, College station.

River Basin Development Authourity of Nigeria (RBDAN), 1979. Constitution of Federal Republic of Nigeria.

Robertson, M.J., Silim, S., Chauhan, Y.S., Ranganathan, R., 2001. Predicting growth and development of pigeonpea: biomass accumulation and partitioning. *Field Crops Research,* 70(2): 89-100. DOI: http://dx.doi.org/10.1016/S0378-4290(01)00125-3.

Robertson, W.K., Hammond, L.C., Johnson, J.T., Boote, K.J., 1980. Effects of Plant-Water Stress on Root Distribution of Corn, Soybeans, and Peanuts in Sandy Soil1. *Agron. J.,* 72(3): 548-550. DOI:10.2134/agronj1980.000219 62007 200030033x.

Rochette, P., Desjardins, R.L., Pattey, E., Lessard, R., 1995. Crop Net Carbon Dioxide Exchange Rate and Radiation Use Efficiency in Soybean. *Agron. J.,* 87(1): 22-28. DOI: 10.2134/agronj1995.00021962008700010005x.

Rockstrom, J., Barron J., Fox P., 2003. Water productivity in rainfed agriculture: challenges and opportunities for smallholder farmers in drought-prone tropical agroecosystems. In: Kijne, J.W., Barker, R., Molden, D. (eds). *Water productivity in agriculture: limits and opportunities for improvement*: 145-162. International Water Management Institute (IWMI), Colombo, Sri Lanka.

Rockstrom, J., 2003. Water for food and nature in drought-prone tropics: vapour shift in rain-fed agriculture. *Philosophical Transactions of the Royal Society of London. Series B: Biological Sciences,* 358(1440): 1997-2009. DOI: 10.1098/rstb.2003.14 00.

Rockstrom, J., Hatibu, N., Oweis, T.Y., Wani, S., Barron, J., Bruggeman, A., Farahani, J., Karlberg, L., Qiang, Z., 2007. Managing Water in Rainfed Agriculture, In: Water for Food, Water for Life: 315-352. Earthscan. London, United Kingdom and International Water Management Institute (IWMI). Colombo, Sri Lanka.

Rodriguez-Diaz, J.A., Weatherhead, E.K., Knox, J.W., Camacho, E., 2007. Climate change impacts on irrigation water requirements in the Guadalquivir river basin in Spain. *Regional Environmental Change,* 7(3): 149-159. DOI: 10.1007/s 10113-007-0035-3.

Rodriguez, D., Sadras, V.O. 2007. The limit to wheat water-use efficiency in eastern Australia. I. Gradients in the radiation environment and atmospheric demand. *Australian Journal of Agricultural Research,* 58(4): 287-302. DOI: http://dx.doi.org/10.1071/AR06135.

Rosegrant, M, Cai X., Cline S., 2002. World Water and Food to 2025. Dealing with Scarcity, International Food Policy Research Institute, Washington DC, USA.

Rosegrant, M.W., Cline, S.A, Li, W., Sulser, T.B., Valmonte-Santos, R.A., 2005. Looking Ahead, Long Term Prospects for Africa's Agricultural Development and Food Security. 2020 Discussion paper 41, International Food Policy Research Institute, 2020 Vision for Food, Agriculture and the Environment, Washington DC, USA.

Roth, G.W., 2013. Soybeans. In. Kirsten, A. (ed.) Agronomy guide, 2013-2014.The Pennsylvania State University: 81-83, USA.

Sadoff, C.W., Whittington, D., 2002. Africa's International Rivers: An Economic Perspective. The World Bank. Washington DC, USA.

Sadok, W., Sinclair, T.R., 2009a. Genetic Variability of Transpiration Response to Vapor Pressure Deficit among Soybean Cultivars. *Crop Sci.,* 49(3): 955-960. DOI: 10.2135/cropsci2008.09.0560.

Sadok, W., Sinclair, T.R., 2009b. Genetic variability of transpiration response to vapor pressure deficit among soybean *(Glycine max (L.) Merr.)* genotypes selected from a recombinant inbred line population. *Field Crops Research,* 113(2): 156-160. DOI: http://dx.doi.org/10.1016/j.fcr.2009.05.002.

Saeki, T., 1960. Interrelationships between leaf amount, light distribution and total photosynthesis in a plant community. Bot. Mag. Tokyo 73: 55-63.

Saha, S., Sehgal, V.K., Nagarajan, S., Pal, M., 2012. Impact of elevated atmospheric CO_2 on radiation utilization and related plant biophysical properties in pigeon pea (*Cajanus cajan L.*). *Agricultural and Forest Meteorology*, 158-159(0): 63-70. DOI: http://dx.doi.org/10.1016/j.agrformet.20 12 .02.003.

Sandana, P., Ramirez, M., Pinochet, D., 2012. Radiation interception and radiation use efficiency of wheat and pea under different P availabilities. *Field Crops Research*, 127(0): 44-50. DOI: http://dx.doi.org/10.1016/j.fcr.2011.11.005.

Santos, J.B.D., Procopio, S.D.O., Silva, A.A.D., Costa, L.C., 2003. Capture and utilization of solar radiation by soybean and bean and weeds. *Bragantia*, 62: 147-153.

Sau, F., Boote, K.J., Ruiz-Nogueira, B., 1999. Evaluation and improvement of CROPGRO-soybean model for a cool environment in Galicia, northwest Spain. *Field Crops Research*, 61(3): 273-291. DOI: http://dx.doi.org /10.1016/S0378-4290(98)00168-3.

Saxton, K.E, Rawls, W.J., Romberger, J.S., Papendick, R.I., 1986. Estimating generalized soil-water characteristics from texture. *Soil Sci. Soc. Am. J.* 50(4): 1031-1036. DOI: 10.1016/0 016-7061(94)90095-7.

Schneider, A., Mathers, A., 1970. Deep ploughing for increased grain sorghum yields under limited irrigation. *Journal of Soil and Water Conservation*, 25(4): 147-150.

Schoffel, E.R.,Volpe, C.A., 2001. Conversion efficiency of photosynthetically active radiation intercepted by soybean for the production of biomass, *Brazilian Journal of Agrometeorology*, Santa Maria, 9(2): 241-249.

Schultz, B., Thatte, C.D., Labhsetwar, V.K., 2005. Irrigation and drainage, Main contributors to global food production. *Irrigation and Drainage*, 54(3): 263-278. DOI: 10.1002/ird.170.

Seckler, D., 1996. The new era of water resource management: From 'dry' to 'wet' water savings, Research Report 1. International Water Management Institute (IWMI), Colombo, Sri Lanka.

Sepaskhah, A.R., Akbari, D., 2005. Deficit Irrigation Planning under Variable Seasonal Rainfall. *Biosystems Engineering*, 92(1): 97-106. DOI: http://dx.doi.or g/10.1016/j.biosystemseng.2005.05.014.

Seyfi, K., Rashidi, M., 2007. Effect of drip irrigation and plastic mulch on crop yield and yield components of cantaloupe. *International Journal of Agriculture and Biology, Leuven*, 2: 247-249.

Sharma, N.P, Damhaup, T., Gilgan-Hunt, E., Grey, D., Okaru, V., Rothberge, D., 1996. African Water Resources: Challenges and Opportunities for Sustainable Development, Technical Paper No. 331. Africa Technical Department Series). The World Bank, Washington DC, USA.

Shiklomanov, I.A., 2000. Appraisal and assessment of world water resources. *Water International*, 25(1): 11-32. DOI: 10.1080/02508060008686794.

Siahpoosh, M.R., Dehghanian, E., 2012. Water Use Efficiency, Transpiration Efficiency, and Uptake Efficiency of Wheat during Drought. *Agron. J.*, 104(5): 1238-1243. DOI: 10.2134/agronj2011.0320.

Siddique, K.H.M., Sedgley, R.H., 1986. Canopy development modifies the water economy of chickpea (*Cicer arietinum L.*) in south-western Australia. *Australian Journal of Agricultural Research*, 37(6): 599-610. DOI: http://dx.doi .org/10.107 1/AR9860599.

Siddique, K.H.M., Loss, S.P., Pritchard, D.L., Regan, K.L., Tennant, D., Jettner, R.L., Wilkinson, D., 1998. Adaptation of lentil (*Lens culinaris Medik.*) to Mediterranean-type environments: effect of time of sowing on growth, yield, and water use. *Australian Journal of Agricultural Research,* 49(4): 613-626. DOI: http://dx.doi.org/10.1071/A97128.

Sincik, M., Candogan, B.N., Demirtas, C., Buyukcangaz, H., Yazgan, S., Goksoy, A.T., 2008. Deficit Irrigation of Soyabean *(Glycine max (L.) Merr.)* in a Sub-humid Climate. *Journal of Agronomy and Crop Science,* 194(3): 200-205. DOI: 10.1111 /j.1439-037X.2008.00307.x.

Sinclair, T.R., Tanner, C.B., Bennett, J.M., 1984. Water-Use Efficiency in Crop Production. *BioScience,* 34(1): 36-40. DOI: 10.2307/1309424.

Sinclair, T.R., Muchow, R.C., 1999. Radiation Use Efficiency. In Donald, L.S. (ed.), *Advances in agronomy* (Vol. Volume 65, 215-265): Academic Press.

Sinclair, T.R., Horie, T., 1989. Leaf Nitrogen, Photosynthesis, and Crop Radiation Use Efficiency: A Review. *Crop Sci.,* 29(1): 90-98. DOI: 10.2135/cropsci1989 .0011183X002900010023x.

Sinclair, T.R., Shiraiwa, T., Hammer, G.L., 1992. Variation in Crop Radiation-Use Efficiency with Increased Diffuse Radiation. *Crop Sci.,* 32(5): 1281-1284. DOI: 10.2135/cropsci1992.0011183X003200050043x.

Sinclair, T.R., Shiraiwa, T., 1993. Soybean Radiation-Use Efficiency as Influenced by Nonuniform Specific Leaf Nitrogen Distribution and Diffuse Radiation. *Crop Sci.,*33(4): 808-812. DOI: 10.2135/cropsci1993.0011183X003300040036x.

Sinclair, T.R., 2000. Model analysis of plant traits leading to prolonged crop survival during severe drought. *Field Crops Research,* 68(3): 211-217. DOI: http://dx.d oi.or g/10.1016/S0378-4290(00)00125-8.

Singer, J.W., Meek, D.W., Sauer, T.J., Prueger, J.H., Hatfield, J.L., 2011. Variability of light interception and radiation use efficiency in maize and soybean. *Field Crops Research,* 121(1): 147-152. DOI: http://dx.doi.org/10.101 6/j.fcr.2010.12.007.

Singh, S.R., Rachie, K.O., Dashiell, K.E. (eds), 1987. Soybeans for the tropics: research, production and utilization. John Wiley & Sons, Chichester, United Kingdom.

Sionit, N., Kramer, P.J., 1977. Effect of Water Stress During Different Stages of Growth of Soybean. *Agron. J.,* 69(2): 274-278. DOI: 10.2134/agronj1977.000 2196200 69 00020018x.

Sivakumar, M.V.K., Shaw, R.H., 1978. Leaf Response to Water Deficits in Soybeans. *Physiologia Plantarum,* 42(1): 134-138. DOI: 10.1111/j.1399-3054.1978.tb0155 3.x.

Smith, A.M., Rakow D.A., 1992. Strategies for reducing water input in woody landscape plantings. *J. Arboricult.*18: 165-170.

Smith, B., 2006. The Farming Handbook, University of KwaZulu-Natal press and Technical Centre for Agricultural and Rural Cooperation CTA: 290-293.

Smith, M., 2000. The Application of Climatic Data for Planning and Management of Sustainable Rainfed and Irrigated Crop Production. *Agricultural and Forest Meteorology,* 103(2): 99-108. DOI: 10.1016/S0168-1923(00)00121-0.

Soltani, A., Sinclair, T.R., 2012. Modelling physiology of crop development, growth and yield: CABI: 117-128.

Soil Conservation Service (SCS), 1991. Soil-Plant-Water Relationships. Irrigation. Section15 Chapter 1. p. 1-1 to 1-56. *In:* National Engineering Handbook, Soil Conservation Service, USDA, Washington DC, USA.

Soil Conservation Service (SCS), 1993. Soil-Plant-Water Relationships. Irrigation. Section15 Chapter 1. p. 1-1 to 1-56. *In:* National Engineering Handbook, Soil Conservation Service, USDA, Washington DC, USA.

Specht, J.E., Elmore, R.W., Eisenhauer, D.E., Klocke, N.W., 1989. Growth stage scheduling criteria for sprinkler-irrigated soybeans. *Irrigation Science,* 10(2): 99-111. DOI: 10.1007/BF00265687.

Steduto, P., Albrizio, R., 2005. Resource use efficiency of field-grown sunflower, sorghum, wheat and chickpea: II. Water use efficiency and comparison with radiation use efficiency. *Agricultural and Forest Meteorology,* 130(3-4): 269-281. DOI: http://dx.doi.org/10.1016/j.agrformet.2005.04.003.

Steduto, P., Hsiao, T., Fereres, E., 2007. On the conservative behavior of biomass water productivity. *Irrigation Science,* 25(3): 189-207. DOI: 10.1007/s00271-007-0064-1.

Steduto, P., Hsiao T.C., Raes, D., Fereres, E., 2009a. AquaCrop The FAO Crop Model to Simulate Yield Response to Water: I. Concepts and Underlying Principles, *Agronomy Journal,* 101(3): 426-437. DOI: 10.2134/agronj2008.0139s.

Steduto, P., Raes, D., Hsiao, T.C., Fereres, E., Heng, L.K., Howell, T.A., Evett, S.R., Rojas-Lara, B.A., Farahani, H.J., Izzi1, G., Oweis, T.Y., Wani, S.P., Hoogeveen, J., Geerts, S., 2009b. Concepts and Applications of AquaCrop: The FAO Crop Water Productivity Model. In: Cao, W., Jeffrey W. White, J.W., Wang, E., Crop Modelling and Decision Support, Springer, 175-191.

Steduto, P., Hsiao T.C., Fereres, E., Raes, D., 2012. Crop yield response to water, FAO Irrigation and Drainage Paper No, 66: 124-131.

Stegman E.C., Hanks R.J., Musick J.T., Watts D.G., 1980 Irrigation water management-adequate or limited water. In: Challenges of the 80's. Proceedings of the ASAE. 2nd National Irrigation Symposium, October.

Stegman, E.C., Schatz, B.G., Gardner, J.C., 1990. Yield sensitivities of short season soybeans to irrigation management. *Irrigation Science,* 11(2): 111-119. DOI: 10. 1007/BF00188447.

Stewart, J.I., Danielson, R.E., Hanks R.J., Jackson, E.B., Haga, R.M., Pruilt, W.O., Franklin, W.T, Riley, J.P., 1997. Optimizing Crop Production Through Control of Water and Salinity Levels in the Soil, PRWG 151-1. Utah State University, Utah Water Laboratory. Logan, Utah, USA.

Stockholm International Water Institute (SIWI), 2001.Water harvesting for upgrading of rainfed agriculture. Problem analysis and research needs. SIWI Report II. Sweden.

Stockle, C.O., and Kemanian, A.R., 2009. Crop Radiation Capture and Use Efficiency: A Framework for Crop Growth Analysis In: Sadras V. and Daniel Calderini D. 2009. Crop Physiology Applications for Genetic Improvement and Agronomy, Academic press: 145-170, USA.

Tefera, H., 2011. Breeding for Promiscuous Soybeans at IITA, Soybean - Molecular Aspects of Breeding, Sudaric, A. (ed.), ISBN: 978-953-307-240-1, InTech, Available from: http://www.intechopen.com/books/Soybean-molecular-aspects-of-breeding/breeding-for-promiscuousSoybeans-at-iita.

Tesfaye, K., Walker, S., Tsubo, M., 2006. Radiation interception and radiation use efficiency of three grain legumes under water deficit conditions in a semi-arid environment. *European Journal of Agronomy,* 25(1): 60-70. DOI: http://dx.doi .org /10.1016/j.eja.2006.04.014.

Thomson, B.D., Siddique, K.H.M., 1997. Grain legume species in low rainfall Mediterranean-type environments II. Canopy development, radiation interception, and dry-matter production. *Field Crops Research*, 54(2-3): 189-1 99. DOI: http://dx.doi.org/10.1016/S0378-4290(97)00048-8.

Todorovic, M., Albrizio, R., Zivotic, L., Saab, M.-T.A., Stockle, C., Steduto, P., 2009. Assessment of AquaCrop, CropSyst, and WOFOST Models in the Simulation of Sunflower Growth under Different Water Regimes. *Agron. J.*, 101(3): 509-521. DOI: 10.2134/agronj2008.0166s.

Tuong, T.P., Pablico, P.P., Yamauchi, M., Confesor, R., Moody, K., 2000. Increasing Water Productivity and Weed Suppression of Wet Seeded Rice: Effect of Water Management and Rice Genotypes. *Experimental Agriculture*, 36(01): 71-89. DOI: doi:null.

Udo, S.O., Aro, T.O., 1999. Global PAR related to global solar radiation for central Nigeria. *Agricultural and Forest Meteorology*, 97(1): 21-31. DOI: http://dx.doi .org /10.1016/S0168-1923(99)00055-6.

Unger, P.W., Jones, O.R., 1981. Effect of Soil Water Content and a Growing Season Straw Mulch on Grain Sorghum. *Soil Sci. Soc. Am. J.*, 45(1): 129-134. DOI: 10.2136 /s ssaj1981.03615995004500010028x.

United Nations Millennium Project (UNMP), 2005a. Halving Hunger: It Can Be Done. Summary version of the report of the Task Force on Hunger, The Earth Institute at Columbia University, New York, USA.

United Nations Millennium Project (UNMP), 2005b. A practical plan to achieve the Millennium Development Goals. Earthscan, New York, USA.

United Nations (UN), 2009. World Population Prospects: The 2008 Revision United Nations (UN) Department of Economic and Social Affairs, Population Division, ESA /P/WP.210. New York, USA.

United Nations World Water Assessment Programme (UN-WWAP), 2006. UN World Water Development Report 2: Water a shared responsibility. United Nations Educational, Scientific and Cultural Organisation (UNESCO) and Berghahn Books. Paris, New York and Oxford, France, USA and United Kingdom.

Vaughan, D.A., Bernard, R.L., Sinclair, J.B., Kunwar, I. K. 1987. Soybean Seed Coat Development. *Crop Sci.*, 27(4): 759-765. doi: 10.2135/cropsci 1987.00 11 183X002700040031x.

Ventura, F., Faber, B., Bali, K., Snyder, R., Spano, D., Duce, P., Schulbach, K., 2001. Model for Estimating Evaporation and Transpiration from Row Crops. *Journal of Irrigation and Drainage Engineering*, 127(6): 339-345. DOI: 10.1061/ (ASCE)0733-9437.

Vinvine, H., Weston V.J., Montgomery R.E., Smyth A.J and Moss R.R., 1954. Progress of soil surveys in South Western Nigeria. Proceedings second Inter-African Soil conference, Leopoldville, Congo.

Vorosmarty, C.J., Green, P., Salisbury, J., Lammers, R.B., 2000. Global Water Resources: Vulnerability from Climate Change and Population Growth. *Science*, 289 (5477): 284-288. DOI: 10.1126/science.289.5477.284.

Vorosmarty, C.J., Lettenmaier, D., Leveque, C., Meybeck, M., Pahl-Wostl, C., Alcamo J., Cosgrove, W., Graßl, H., Hoff, H., Kabat, P., Lansigan, F, Lawford, R., Naiman, R., 2004. Humans transforming the global water system. EOS, Transactions, *American Geophysical Union*, 85(48): 509-514. DOI: 10.1029/2004 EO480001.

Wang, X., Zhao, C., Chen, H., 1993. Study on Water Saving Agriculture and Water Saving Techniques, Meteorology Press, Beijing, China.

Wang, D., Shannon, M.C., Grieve, C.M., 2001. Salinity reduces radiation absorption and use efficiency in Soybean. *Field Crops Research,* 69(3): 267-277. DOI: http://dx.doi.org/10.1016/S0378-4290(00)00154-4.

Wani, S.P., Sreedevi, T.K., Rockstrom, J., Ramakrishna, Y.S., 2009. Rainfed Agriculture-past trends and future prospects In: Suhas, P., Wani, J.R., Theib, O. 2009. Rainfed Agriculture, Unlocking the Potentials, pp 1-32 Comprehensive assessment of water management in agriculture, CAB Books, United Kingdom.

Waraich, E.A., Ahmad, R., Ashraf, M.Y., Saifullah, Ahmad, M., 2011. Improving agricultural water use efficiency by nutrient management in crop plants. *Acta Agriculturae Scandinavica, Section B- Soil & Plant Science,* 61(4): 291-304. DOI: 10.1080/09064710.2010.491954.

Wardlaw, I.F., 1990. Tansley Review No. 27. The control of carbon partitioning in plants. *New phytologist,* 116(3): 341-381.

Wells, R., Burton, J.W., Kilen, T.C., 1993. Soybean Growth and Light Interception: Response to Differing Leaf and Stem Morphology. *Crop Sci.,* 33(3): 520-524. DOI: 10.2135/cropsci1993.0011183X003300030020x.

Westgate, M.E., Peterson, C.M., 1993. Flower and Pod Development in Water-Deficient Soybeans (Glycine max (L.) Merr.). *Journal of Experimental Botany,* 44(1): 109-117. DOI: 10.1093/jxb/44.1.109.

Westgate, M.E., Whitgham, K., Purcell, L., 2004. Soybean. In:Editor-In-Chief, Wringley, C. 2004. Encyclopaedia of grain science, Elsevier, 146-155.

White, J.W., Catillo, J.A., 1990. Studies at CIAT on mechanisms of drought tolerance in beans. In: White, J.W., Hoogenboom, G., Ibarra, F., Sing, S.P. (eds). Research on Drought tolerance in Common Bean. 146-151. Centro Internacional de Agricultura Tropical. Cali, Colombia.

Winkel, T., Renno, J.F., Payne, W.A., 1997. Effect of the timing of water deficit on growth, phenology and yield of pearl millet *(Pennisetum glaucum (L.) R. Br.)* grown in Sahelian conditions. *Journal of Experimental Botany,* 48(5): 1001-1009. DOI: 10.1093/jxb/48.5.1001.

Wofford, T.J., Allen, F.L., 1982. Variation in Leaflet Orientation among Soybean Cultivars. *Crop Sci.,* 22(5): 999-1004. DOI: 10.2135/cropsci19 82.0011183X00 2200050025x.

World Bank (WB), 2000. Spurring Agricultural and Rural Development. In: Can Africa Claim the 21st Century? Washington DC, USA.

World Bank (WB), 2005. Agricultural Growth for the Poor: An Agenda for Development World Bank, Washington DC, USA.

World Population Prospects (WPP), 2012. The 2012 Revision (Medium variant). Available at http://esa.un.org/unpd. Accessed on 29 January, 2015.

Yaklich, R.W., Vigil, E.L., Wergin, W.P., 1986. Pore Development and Seed Coat Permeability in Soybean1. *Crop Sci.,* 26(3): 616-624. DOI: 10.2135/cropsci1986.0 011183X002600030041x.

Yang, H., Zehnder, A., 2007. Virtual water: An unfolding concept in integrated water resources management, *Water Resources Research,* 43, 10, doi: 10.1029/2 007WR00 6048.

Zarea, M.J., Ghalavand, A., Daneshian, J., 2005. Effect of planting patterns of sunflower on yield and extinction coefficient, *Agron. for Sustain.,* Dev., 25(4): 513-518.

Zeleke, K.T., Luckett, D., Cowley, R., 2011. Calibration and Testing of the FAO AquaCrop Model for Canola. *Agron. J.,* 103(6): 1610-1618. DOI:10.2134/agronj2011.0150.

Zhang, H., Oweis, T., Garabet, S., Pala, M., 1998. Water-use efficiency and transpiration efficiency of wheat under rain-fed conditions and supplemental irrigation in a Mediterranean-type environment. *Plant and Soil,* 201(2): 295-305. DOI: 10.1023/A:1004328004860.

Zhang, H., Owie T., 1999. Water-yield relations and optima irrigation scheduling of wheat in the Mediterranean region. *Agricultural Water Management,* 38(3): 195-211. DOI: 10.1016/S0378-3774(98)00069-9.

Zhang, L., Wang, J., Rozelle, S., 2008. Development of Groundwater Markets in China: A Glimpse into Progress to Date. *World Development,* 36(4): 706-726.

Zhang, X., Cheng, W., Liu, E., 1992. Experimental study on the effects of field mulching on water saving. In: Xu, Y. (ed.). Study on water-saving agriculture: 200-203. Science Press, Beijing, China.

Ziska, L.H., Hall, A.E., 1983. Seed yields and water use of cowpeas *(Vigna unguiculata (L.) Walp.)* subjected to planned-water-deficit irrigation. *Irrigation Science,* 3(4): 237-245.

Zur, B., Jones, J.N., 1981. A model for the water relations, photosynthesis and expansive growth of crops. *Water Res. Res.* 17(2): 311-320. DOI: 10.1029/WR01 7i002p00311.

Zwart, S.J., Bastiaanseen, W.G.M., 2004. Review of measured crop water productivity values for irrigated wheat, rice, cotton and maize. *Agricultural Water Management,* 69(2): 115-133. DOI: 10.1016 /j.agwat.2004.04.007.

Appendices

Appendix A. List of symbols

Symbol	Description	Unit
A^i	Bras and Corodova moisture stress sensitivity index for the growth stage i	-
B	Growth coefficient	-
B	Biomass	t ha^{-1}
BD	Soil bund	-
Bi	growth stage weighing coefficient	-
Bi	Daily above ground biomass	t ha^{-1}
CC	Canopy cover	%
CC_x	Maximum canopy cover	%
cc_o	Initial canopy cover	%
CDC	Canopy decline coefficient	day^{-1}
CDC_{adj}	Adjusted canopy decline coefficient	day^{-1}
CF	carbon dioxide photosynthate value	
C_f	Proportion of chaff in the dry biomass at harvest	%
CGC	Canopy growth coefficient	-
CO_2	Carbon dioxide	ppm
CV	Coefficient of variation	-
CWP	Crop water productivity	kg ha^{-1} mm^{-1}
d	Willmontt's index of agreement	-
DAB	Dry above ground biomass	t ha^{-1}
DAP	Day after planting	day
DD	Drought day	-
D_1	Limiting soil water deficit for optimum growth	mm
DOY	Day of the year	day
dZ	Rate of deeping of root	Mm day^{-1}
E	Evaporation of water from soil	mm
e	Conversion efficiency	G MJ^{-1}
E_a	Actual evaporation of water from cropped soil	mm
EF	Nash-Sutcliff efficient coefficient	-
E_p	Pan evaporation from class A pan evaporator	mm
ET_a	Seasonal actual evapotranspiration or actual crop evapotranspiration from moisture stressed treatment at growth stage i	mm year^{-1}
ET_{ai}	Actual crop evapotranspiration from moisture stressed treatment at growth stage 'i'	mm day^{-1}
ET_i	actual evapotranspiration for stage i	mm day^{-1}
ET_{mi}, ET_m	crop evapotranspiration from non-stressed treatment at growth stage "i"	mm day^{-1}
ET_m	Maximum evapotranspiration	mm day^{-1}
ET_o	Reference evapotranspiration	mm day^{-1}
ET_{oi}	Reference evapotranspiration for a specific day	mm day^{-1}
ET_p	Potential evapotranspiration for stage i	mm day^{-1}
ET_p	Seasonal potential transpiration	mm year^{-1}
ET_p	Potential evapotranspiration	mm day^{-1}
ET_x	Maximum transpiration	Mm
EU	Emission uniformity	%
E_x	Evaporation from bare soil	Mm day^{-1}
F	Fraction of intercepted radiation	-
FC	Field capacity	m^3 m^{-3}

f_e	Proportion of PAR in SR	-
F_i	Fraction of Photosynthetically Active Radiation	-
fIPAR	Fraction of Photosynthetically Active Radiation	-
f_{sink}	Crop sink strength coefficient	-
G	Average rate of yield loss	$g\ m^{-2}\ year^{-1}$
GDD	Growing degree days	
GLA	Green leaf area	$m^2\ m^{-2}$
H	Height of plant	cm
HI	Harvest index	%
HI_o	Reference harvest index	%
I	Irrigation depth	mm
I	growth stage	-
IPAR	Intercepted Photosynthetically Active Radiation	$MJ\ m^{-2}$
IWP	Irrigation water productivity	$kg\ ha^{-1}\ mm^{-1}$
K	Growth constant	-
K_c	Crop coefficient	-
k_{cend}	Crop coefficient at end of the season	-
K_{cmid}	Crop coefficient at mid season	-
k_s	Water stress coefficient	-
k_{sat}	Saturated hydraulic conductivity	$mm\ day^{-1}$
K_{sexp}	Water stress coefficient for canopy expansion	-
ksi	Water stress coefficient for stomata closure	-
Ky	Crop response factor	-
K_y	Stewart's moisture stress yield reduction coefficient	-
LAI	Leaf area index	$m^2\ m^{-2}$
MAE	Mean Absolute Error	-
ML	Mulch	-
MLBD	Mulch plus soil bund	-
N	Number of irrigation/observation	-
N	Number of growth stages	-
N	Shape factor of a function	-
NC	Conventional practice	-
NRMSE	Normalised root means square error	-
p	Fractional depletion of total available water	-
p	market price/kg of the crop	$
PAR_a	Photosynthetically Active Radiation (above)	$\mu\ mol\ m^{-2}\ m^{-1}$
PAR_b	Photosynthetically Active Radiation (below)	$\mu\ mol\ m^{-2}\ m^{-1}$
PFD	Photon flux density	$\mu mol\ m^2\ s^{-1}$
PG	photosynthate assimilation rate	
PH	Plant height	cm
PWP	Permanent wilting point	$m^3\ m^{-3}$
Q_{av}	Average emitters discharge	ls^{-1}
Q_x	Discharge from emitters	ls^{-1}
R	Rainfall	mm
r^2	Coefficient of determination	-
RCBD	Randomised complete block design	-
REW	Readily evaporable water	mm
RMSE	Root means square error	%
RUE	Radiation use efficiency	$g\ MJ^{-1}$
S_d	Proportion of the seed in the dry biomass at harvest	%
SEP_i	Seasonal soil water evaporation during irrigation seasons	mm
SEP_r	Seasonal soil water evaporation during rainy season	mm
Si	Sink term at specific depth	-
SR	Solar radiation	$W\ m^{-2}$
S_t	Proportion of stems in dry biomass at harvest	%
STP_i	Seasonal transpiration during irrigation seasons	mm

STP_r	Seasonal transpiration during rainy season	mm
SWS	Soil water storage	mm
SWU_i	Seasonal water use during irrigation seasons	mm
SWU_r	Seasonal water use during rainy seasons	mm
T	Cumulative seasonal transpiration	mm year^{-1}
T_a	Actual transpiration	mm day^{-1}
TAW	Total available water	mm
T_i	Actual evapotranspiration for stage i	mm year^{-1}
T_i	Transpiration for a particular day	Mm day^{-1}
TIPAR	Total Intercepted Photosynthetically Active Radiation	MJ m^{-2}
T_m	Maximum transpiration	mm day^{-1}
T_n	Minimum air temperature	^{o}C
T_p	Potential transpiration	mm year^{-1}
T_r	Canopy transpiration	mm
TR	Tied ridge	-
TRBD	Tied ridge plus soil bund	-
TRML	Tied ridge plus mulch	-
T_x	Maximum air temperature	^{o}C
U_s	Uniformity coefficient	%
W_{sm}	Available water storage	mm
WP	Water productivity for seed/grain under irrigation conditions	kg ha^{-1} mm^{-1}
WP*	Normalised water productivity for ET_o and CO_2	Mm day^{-1}
$WP_{biomass}$	Water productivity for biomass production under rainfed conditions	kg ha^{-1} mm^{-1}
WP_{seed}	Water productivity for seed/grain under rainfed conditions	kg ha^{-1} mm^{-1}
$WP_{economic}$	Economic water productivity	kg ha^{-1} mm^{-1}
W_{si}	Soil water storage at the end of stage i	mm
WUE	Water Use Efficiency	kg ha^{-1} mm^{-1}
Y	Actual or relative yield	gm^{-2}, %
Y_a	Actual yield	(t ha^{-1})
Y_m	Maximum or potential yield	g m^{-2}
Y_x	Maximum yield	t ha^{-1}
Z	Effective rooting depth	mm
Z_{ini}	Depth at sowing	mm
Zx	Maximum effective rooting depth	mm
λi	growth state weighing coefficient	mm day^{-1}
Λ	Fitted exponent, Jensen's moisture stress sensitivity index	-
Λ	leaf extinction coefficient	-
δ	Minhas's moisture stress sensitivity index	-
τ	Drainage coefficient	-

Appendix B. Acronyms

Acronym	Description
ANOVA	Analysis of variance
AQUASTAT	FAO information data bank
CROPGRO	Crop grow model
CSD	Commission on Sustainable Development
CWP	Crop Water Productivity
DAP	Days after planting
DD	Drought day
DI	Deficit irrigation
DOY	Day of the year
FAO	Food and Agriculture Organization of the United Nations
FAOSTAT	Food and Agriculture Organization of the United Nations statistics
FI	Full irrigation
FMEN	Federal Ministry of Environment, Nigeria
GDP	Gross Domestic product
ICID	International Commission on Irrigation and Drainage
IITA	International Institute for Tropical Agriculture
ITCZ	Inter-tropical Convergence Zone
MDG	Millennium Development Goals
NEEDS	National Economic Empowerment and Development Strategies
NINCID	Nigeria National Commission for Irrigation and Drainage
NMDG	National Millennium Development Goals
NPCN	National Population Commission of Nigeria
OORBA	Ogun-Osun River Basin Authority
RBRDA	River Basin and Rural Development Authority
SCS	Soil Conservation Service
SEEDS	State Economic Empowerment and Development Strategies
SEP_r	Evaporation under rainfed conditions
SH	Sub humid
SIWI	Stockholm International Water Institute
SSA	Sub-Saharan Africa
TRFOAU	Teaching and Research farms of Obafemi Awolowo University
UN	United Nations
UNWWAP	United Nations World Water Assessment Programme
USDA	United State Department of Agriculture
VPD	Vapour pressure deficit
WARM	Water Accounting Rice Model
WB	World bank
WOFOST	World Food Study

Appendix C. Meteorological variables logged at intervals of 10 minutes at the experimental site from 2011 - 2014

		Air temperature (°C)			Relative humidity (%)			Global solar radiation (W m⁻²)		Actual vapour pressure (KPa)		
2011		Max	Min	Mean	Max	Min	Mean	Max	Mean	Max	Min	Mean
	Apr	34.8	20.8	26.5 (3.2)	95.2	29.2	77.0 (13.0)	984	100 (233)	3.22	1.96	2.65 (0.15)
	May	32.3	20.7	25.4 (2.5)	94.7	43.5	81.2 (11.8)	1077	172 (253)	3.10	2.33	2.76 (0.10)
	Jun	31.0	20.1	25.4 (2.5)	94.5	28.2	82.4 (10.6)	1089	156 (241)	3.08	2.18	2.66 (0.12)
	Jul	29.4	20.1	23.8 (1.9)	94.9	55.4	84.5 (7.80)	957	125 (182)	2.96	2.21	2.54 (0.10)
	Aug	39.5	20.1	23.7 (1.8)	95.2	55.3	86.3 (8.00)	1058	117 (201)	2.85	2.18	2.53 (0.10)
	Sep	31.7	19.9	24.5 (2.2)	95.3	52.2	52.2 (9.40)	1228	144 (211)	2.96	2.06	2.59 (0.09)
	Oct	31.9	20.0	26.5 (3.0)	95.3	48.4	80.3 (10.1)	1080	164 (245)	2.98	2.05	2.58 (0.12)
	Nov	34.0	19.2	28.1 (3.6)	94.5	19.2	72.4 (15.8)	982	202 (286)	2.93	1.50	2.57 (0.23)
	Dec	35.2	15.0	25.7 (4.6)	95.7	15.0	58.2 (22.7)	819	193 (264)	2.88	0.00	1.79 (0.34)
2012	Jan	35.1	15.2	25.7 (4.3)	95.3	11.2	57.4 (19.9)	693	171 (223)	2.89	0.57	1.85 (0.28)
	Feb	34.8	20.3	26.6 (3.2)	94.2	15.5	73.9 (16.1)	775	178 (240)	2.92	0.63	2.50 (0.22)
	Mar	36.4	20.0	27.8 (3.5)	94.3	11.2	69.5 (18.8)	847	216 (274)	3.12	0.64	2.52 (0.33)
	Apr	35.3	21.1	27.0 (3.1)	94.8	34.2	76.5 (15.1)	989	192 (266)	3.11	1.93	2.66 (0.15)
	May	33.8	19.5	25.9 (3.0)	94.8	42.6	80.1 (12.3)	1084	177 (247)	2.97	2.08	2.64 (0.14)
	Jun	32.3	20.1	24.9 (2.4)	95.1	51.0	83.6 (10.3)	981	162 (222)	2.97	2.11	2.60 (0.10)
	Jul	32.3	20.0	23.8 (1.8)	98.0	51.0	86.9 (7.30)	973	120 (175)	2.86	2.21	2.56 (0.07)
	Aug	29.3	20.0	23.2 (1.6)	95.1	51.0	87.4 (6.90)	965	101 (145)	2.80	2.05	2.48 (0.08)
	Sept	29.6	20.8	24.0 (1.8)	95.4	59.9	87.1 (8.00)	1003	117 (177)	2.85	2.34	2.58 (0.06)
	Oct	31.0	20.2	24.7 (2.6)	95.6	52.3	54.4 (10.7)	1003	160 (236)	2.97	2.12	2.51 (0.06)
	Nov	37.7	20.5	26.1 (2.8)	95.3	39.9	80.0 (12.9)	1008	187 (272)	3.10	1.84	2.65 (0.16)
	Dec	34.9	16.6	26.4 (3.7)	95.3	14.3	66.5 (19.7)	1817	185 (259)	2.32	0.32	2.32 (0.32)

2013	Jan	35.5	15.0	26.0(3.7)	94.1	14.5	55.7(26.5)	814	188(264)	4.96	0.21	2.33 (0.33)
	Feb	41.0	18.0	27.5 (3.7)	94.3	10.1	66.0(18.6)	904	161(234)	7.27	0.27	2.73 (0.58)
	Mar	34.5	21.3	27.2(3.4)	94.1	42.4	76.4(14.5)	810	128(219)	5.11	1.18	3.05 (0.11)
	Apr	34.8	21.7	25.8(3.7)	94.5	40.4	78.5(13.7)	1003	190(266)	5.13	1.06	2.98 (0.17)
	May	37.0	20.8	26.1(2.7)	95.6	15.6	81.5(12.9)	985	180(245)	5.97	0.38	2.91 (0.12)
	June	32.6	20.2	25.1(2.2)	96.5	46.5	84.1(10.5)	940	153(208)	4.91	1.32	2.84 (0.09)
	July	30.1	19.6	23.8(2.0)	100	56.4	89.9(10.5)	961	129(188)	4.21	1.41	2.73 (0.12)
	Aug	29.6	20.2	23.5(1.8)	100	49.5	89.9(11.6)	589	48.8(112)	4.14	2.37	3.06 (0.42)
	Sept	31.3	11.5	24.3(2.3)	100	52.2	92.6(13.9)	903	94.2(177)	4.58	1.36	3.29(0.44)
	Oct	32.3	20.6	24.8(2.5)	100	50.9	92.8(25.2)	898	113(170)	4.84	2.43	3.45(0.34)
	Nov	33.5	20.5	26.3(2.8)	100	37.9	87.2(22.2)	793	97.8(202)	5.18	2.41	3.75(0.37)
	Dec	33.1	16.7	25.9(3.3)	100	20.3	78.6(23.5)	837	179(250)	5.80	0.39	3.10(0.32)
2014	Jan	35.4	18.1	26.4(3.2)	100	15.1	81.3(25.2)	841	152(219)	6.54	0.40	3.33(0.35)
	Feb	36.3	19.7	27.5(3.7)	100	13.5	68.8(25.4)	798	166(229)	6.80	0.31	3.27(0.26)

Note: Standard deviations in parenthesis, 1 MJ m^{-2} day^{-1} = 11.574 W m^{-2}

Appendix D. Average soil temperatures in the upper 30 and lower 30 cm of the soil profile under different conservation practices and the conventional method in the rainy seasons

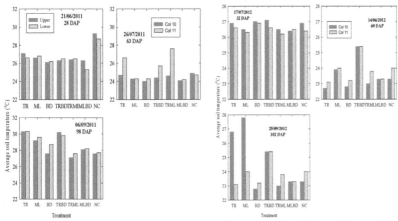

Figure D.1. Average soil temperatures in the upper 30 and lower 30 cm of the soil profile under different conservation practices and the conventional method in the rainy season

Appendix E. Analysis of variance (ANOVA) of soil water storage

Appendix E.1. Statistical analysis of the soil water storage in 2011

26 DAP

Source	DF	Sum of squares	Mean square	F value	Pr > F
Model	6	1895	316	3.79	0.02
Error	14	1167	83.3		
Corrected total	20	3062			
R-square	Coeff var	Root MSE	SMS mean		
0.618886	10.8	9.13	84.6		
Source	DF	Type III SS	Mean square	F value	Pr > F
Treatment	6	1895	316	3.79	0.02

38 DAP

Source	DF	Sum of squares	Mean square	F value	Pr > F
Model	6	1299	216	4.86	0.01
Error	14	623	44.5		
Corrected total	20	1922			
R-square	Coeff var	Root MSE	SWS mean		
0.675685	6.81	6.67	98.0		
Source	DF	Type III SS	Mean square	F value	Pr > F
Treatment	6	1299	216	4.86	0.01

61 DAP

Source	DF	Sum of squares	Mean square	F value	Pr > F
Model	6	1357	226	5.07	0.0059
Error	14	625	44.6		
Corrected total	20	1981			
R-square	Coeff var	Root MSE	SWS mean		
0.684709	7.93	6.68	84.2		
Source	DF	Type III SS	Mean square	F value	Pr > F
Treatment		1357	226	5.07	0.01

97 DAP

Source	DF	Sum of squares	Mean square	F value	Pr > F
Model	6	1220	203	1.37	0.29
Error	14	2073	148		
Corrected total	20	3292			
R-square	Coeff var	Root MSE	SWS mean		
0.370373	14.1	12.1	86.4		
Source	DF	Type III SS	Mean square	F value	Pr > F
Treatment	6	1220	203	1.37	0.29

117 DAP

Source	DF	Sum of squares	Mean square	F value	Pr > F
Model	6				
Error	14				
Corrected total	20				
R-square	Coeff var	Root MSE	SWS mean		
0.370373					
Source	DF	Type III SS	Mean square	F value	Pr > F
Treatment	6				

Appendix E.2. Statistical analysis of the soil water storage in 2012

28 DAP

Source	DF	Sum of squares	Mean square	F value	Pr > F
Model	6	506	84.3	1.34	0.30
Error	14	881	62.9		
Corrected total	20	1386			
R-square	Coeff var	Root MSE	SWS mean		
0.364742	6.71	7.93	118		
Source	DF	Type III SS	Mean square	F value	Pr > F
Treatment	6	506	84.3	1.34	0.30

43 DAP

Source	DF	Sum of squares	Mean square	F value	Pr > F
Model	6	.1277	213	6.33	0.00
Error	14	471	33.6		
Corrected total	20	1748			
R-square	Coeff var	Root MSE	SWS mean		
0.730671	5.52	5.80	105		
Source	DF	Type III SS	Mean square	F value	Pr > F
Treatment	6	1277	213	6.33	0.00

58 DAP

Source	DF	Sum of squares	Mean square	F value	Pr > F
Model	6	381	63.5	0.83	0.57
Error	14	1069	76.4		
Corrected total	20	1450			
R-square	Coeff var	Root MSE	SWS mean		
0.262617	8.55	8.74	102		
Source	DF	Type III SS	Mean square	F value	Pr > F
Treatment		381	63.5	0.83	0.57

98 DAP

Source	DF	Sum of squares	Mean square	F value	Pr > F
Model	6	2753	459	6.30	0.00
Error	14	1020	72.9		
Corrected total	20	3773	72.9		
R-square	Coeff var	Root MSE	SWS mean		
0.729578	6.87	8.54			
Source	DF	Type III SS	Mean square	F value	Pr > F
Treatment	6	2753	459	6.30	0.00

117 DAP

Source	DF	Sum of squares	Mean square	F value	Pr > F
Model	6	1395	232	1.60	0.22
Error	14	2040	146		
Corrected total	20	3435			
R-square	Coeff var	Root MSE	SWS mean		
0.406068	10.9	12.1	111		
Source	DF	Type III SS	Mean square	F value	Pr > F
Treatment	6	1395	233	1.60	0.22

Appendix E.3. Statistical analysis of the (a) plant heights and (b) number of leaves in 2011 under rainfed conditions

(a)   VE-V2

Source	DF	Sum of squares	Mean square	F value	Pr > F
Model	6	16.1	2.68	6.61	0.0005
Error	21	8.50	0.41		
Corrected total	27	24.54			
R-square	Coeff var	Root MSE	PH mean		
0.65	3.31	0.64	19.2		
Source	DF	Type III SS	Mean square	F value	Pr > F
Treatment	6	16.1	2.68	6.61	0.0005

Flowering

Source	DF	Sum of squares	Mean square	F value	Pr > F
Model	6	45.6	7.76	1.92	0.124
Error	21	84.8	4.04		
Corrected total	27	131			
R-square	Coeff var	Root MSE	PH mean		
0.35	6.12	2.01	32.9		
Source	DF	Type III SS	Mean square	F value	Pr > F
Treatment	6	45.6	7.76	1.92	0.124

Pod initiation

Source	DF	Sum of squares	Mean square	F value	Pr > F
Model	6	390	65.2	2.93	0.0308
Error	21	467	22.2		
Corrected total	27	858			
R-square	Coeff var	Root MSE	PH Mean		
0.46	8.76	4.72	53.8		
Source	DF	Type III SS	Mean square	F value	Pr > F
Treatment	6	390	65.2	2.93	0.0308

Seed filling

Source	DF	Sum of squares	Mean square	F value	Pr > F
Model	6	840	140	504	0.001
Error	21	5.83	0.28		
Corrected total	27				
R-square	Coeff var	Root MSE	PH Mean		
0.99	0.95	0.53	55.4		
Source	DF	Type III SS	Mean square	F value	Pr > F
Treatment	6	840	140	504	0.001

Maturity

Source	DF	Sum of squares	Mean square	F value	Pr > F
Model	6	845	141	468	0.001
Error	21	6.32	0.30		
Corrected total	27	852			
R-square	Coeff var	Root MSE	PH Mean		
0.99	0.99	0.54	55.4		
Source	DF	Type III SS	Mean square	F value	Pr > F
	6	845	141	468	0.001

(b) VE-V2

Source	DF	Sum of squares	Mean square	F value	Pr > F
Model	6	25.8	4.30	8.18	0.0001
Error	21	11.4	0.53		
Corrected total	27	36.9			
R-square	Coeff var	Root MSE	NL mean		
0.70	3.52	0.73	20.6		
Source	DF	Type III SS	Mean square	F value	Pr > F
Treatment	6	25.8	4.30	8.18	0.0001

Flowering

Source	DF	Sum of squares	Mean square	F value	Pr > F
Model	6	1161	194	5.10	0.0023
Error	21	797	37.9		
Corrected total	27	1958			
R-square	Coeff var	Root MSE	NL mean		
0.59	11.9	6.16	51.4		
Source	DF	Type III SS	Mean square	F value	Pr > F
Treatment	6	1161	194	5.10	0.0023

Pod initiation

Source	DF	Sum of squares	Mean square	F value	Pr > F
Model	6	40311	6719	5.65	0.0013
Error	21	24988	1189		
Corrected total	27	65300			
R-square	Coeff var	Root MSE	NL mean		
0.62	13.2	34.5	262		
Source	DF	Type III SS	Mean square	F value	Pr > F
Treatment	6	40311	6719	5.65	0.0013

Seed filling

Source	DF	Sum of squares	Mean square	F value	Pr > F
Model	6	52200	8700	173334	0.0001
Error	21	10.5	0.50		
Corrected total	27	52210			
R-square	Coeff var	Root MSE	NL mean		
0.99	0.26	0.71	273		
Source	DF	Type III SS	Mean square	F value	Pr > F
Treatment	6	52200	8700	173334	0.0001

Maturity

Source	DF	Sum of squares	Mean square	F value	Pr > F
Model	6	50812	8469	12480	0.0001
Error	21	14.3	0.68		
Corrected total	27	50827			
R-square	Coeff var	Root MSE	NL mean		
0.99	0.32	0.83	261		
Source	DF	Type III SS	Mean square	F value	Pr > F
Treatment	6	50812	8469	12480	0.0001

Appendix E.4. Statistical analysis of the (a) plant heights and (b) number of leaves in 2012 under rainfed conditions

(a)   plant height

VE-V2

Source	DF	Sum of squares	Mean square	F value	Pr > F
Model	6	138	22.9	81	<.0001
Error	14	3.94	0.28		
Corrected total	20	142			
R-square	Coeff var	Root MSE	PH Mean		
0.97	2.82	0.53	18.8		
Source	DF	Type III SS	Mean square	F value	Pr > F
Treatment	6	138	22.9	81	<.0001

Flowering

Source	DF	Sum of squares	Mean square	F value	Pr > F
Model	6	333	55.4	345	<.0001
Error	14	2.24	0.16		
Corrected total	20	335			
R-square	Coeff var	Root MSE	PH Mean		
0.99	1.17	0.40	34.2		
Source	DF	Type III SS	Mean square	F value	Pr > F
Treatment	6	333	55.4	345	<.0001

Pod initiation

Source	DF	Sum of squares	Mean square	F value	Pr > F
Model	6	131	21.9	1.11	0.41
Error	14	277	19.8		
Corrected total	20	409			
R-square	Coeff var	Root MSE	PH Mean		
0.32	8.71	4.45	51.1		
Source	DF	Type III SS	Mean square	F value	Pr > F
Treatment	6	131	21.9	1.11	0.41

Seed filling

Source	DF	Sum of squares	Mean square	F value	Pr > F
Model	6	756	126	156	<.0001
Error	14	11.3	0.81		
Corrected total	20	766			
R-square	Coeff var	Root MSE	PH Mean		
0.99	1.23	0.90	72.9		
Source	DF	Type III SS	Mean square	F value	Pr > F
Treatment	6	756	126	156	<.0001

Maturity

Source	DF	Sum of squares	Mean square	F value	Pr > F
Model	6	724	121	93.8	<.0001
Error	14	18	1.29		
Corrected total	20	742			
R-square	Coeff var	Root MSE	PH Mean		
0.98	1.56	1.13	72.9		
Source	DF	Type III SS	Mean square	F value	Pr > F
Treatment	6	724	121	93.8	<.0001

Number of leaves

VE-V2

Source	DF	Sum of squares	Mean square	F value	Pr > F
Model	6	53.9	8.98	11.8	<.0001
Error	14	10.7	0.76		
Corrected total	20	65.6			
R-square	Coeff var	Root MSE	NL mean		
0.83	4.40	0.87	19.9		
Source	DF	Type III SS	Mean square	F value	Pr > F
Treatment	6	53.9	8.98	11.8	<.0001

Flowering

Source	DF	Sum of squares	Mean square	F value	Pr > F
Model	6	676	113	215	<.0001
Error	14	7.33	0.52		
Corrected total	20	683			
R-square	Coeff var	Root MSE	NL mean		
0.99	1.19	0.72	60.5		
Source	DF	Type III SS	Mean square	F value	Pr > F
Treatment	6	676	113	215	<.0001

Pod initiation

Source	DF	Sum of squares	Mean square	F value	Pr > F
Model	6	4104	684	4.64	0.0085
Error	14	2063	147		
Corrected total	20	6167			
R-square	Coeff var	Root MSE	NL mean		
0.67	7.73	12.1	157		
Source	DF	Type III SS	Mean square	F value	Pr > F
Treatment	6	4104	684	4.64	0.0085

Seed filling

Source	DF	Sum of squares	Mean square	F value	Pr > F
Model	6	17435	2906	1695	<.0001
Error	14	24			
Corrected total	20	17459			
R-square	Coeff var	Root MSE	NL mean		
0.99	0.44	1.31	300		
Source	DF	Type III SS	Mean square	F value	Pr > F
Treatment	6	17435	2906	1695	<.0001

Appendix F. Analysis of variance (ANOVA) of the leaf area indices in (a) 2011 and (b) 2012 under rainfed conditions

(a)

VE-V2

Source	DF	Sum of squares	Mean square	F value	Pr > F
Model	6	0.09	0.02	0.87	0.53
Error	21	0.39	0.02		
Corrected total	27	0.49			
R-square	Coeff var	Root MSE	LAI Mean		
0.20	24.8	0.14	0.55		
Source	DF	Type III SS	Mean square	F value	Pr > F
Treatment	6	0.09	0.02	0.87	0.53

Flowering

Source	DF	Sum of squares	Mean square	F value	Pr > F
Model	6	0.12	0.02	0.59	0.74
Error	21	0.70	0.03		
Corrected total	27	0.82			
R-square	Coeff var	Root MSE	LAI Mean		
0.14	23.3	0.18	0.78		
Source	DF	Type III SS	Mean square	F value	Pr > F
Treatment	6	0.12	0.02	0.59	0.74

Pod initiation

Source	DF	Sum of squares	Mean square	F value	Pr > F
Model	6	2.93	0.49	3.51	0.015
Error	21	2.92	0.14		
Corrected total	27	5.86			
R-square	Coeff var	Root MSE	LAI Mean		
0.51	19.9	0.37	1.87		
Source	DF	Type III SS	Mean square	F value	Pr > F
Treatment	6	2.93	0.49	3.51	0.015

Seed filling

Source	DF	Sum of squares	Mean square	F value	Pr > F
Model	6	9.69	1.62	1.27	0.31
Error	21	26.6	1.27		
Corrected total	27				
R-square	Coeff var	Root MSE	LAI Mean		
0.27	28.2	1.13	4.00		
Source	DF	Type III SS	Mean square	F value	Pr > F
Treatment	6	9.69	1.62	1.27	0.31

Maturity

Source	DF	Sum of squares	Mean square	F value	Pr > F
Model	6	2.05	0.34	12.1	<.0001
Error	21	0.59	0.03		
Corrected total	27	2.65			
R-square	Coeff var	Root MSE	LAI Mean		
0.78	9.72	0.17	1.73		
Source	DF	Type III SS	Mean square	F value	Pr > F
Treatment	6	2.05	0.34	12.1	<.0001

(b)

VE-V2

Source	DF	Sum of squares	Mean square	F value	Pr > F
Model	6	0.04	0.01	217	<.0001
Error	14	0.00	0.00		
Corrected total	20	0.04			
R-square	Coeff var	Root MSE	LAI Mean		
0.99	2.18	0.06	0.27		
Source	DF	Type III SS	Mean square	F value	Pr > F
Treatment	6	0.04	0.01	217	<.0001

Flowering

Source	DF	Sum of squares	Mean square	F value	Pr > F
Model	6	1.11	0.19	3.93	0.02
Error	14	0.67	0.05		
Corrected total	20	1.77			
R-square	Coeff var	Root MSE	LAI Mean		
0.63	21.4	0.22	1.01		
Source	DF	Type III SS	Mean square	F value	Pr > F
Treatment	6	1.11	0.19	3.93	0.02

Pod initiation

Source	DF	Sum of squares	Mean square	F value	Pr > F
Model	6	2.13	0.36	3551	<.0001
Error	14	0.00	0.00		
Corrected total	21	2.13			
R-square	Coeff var	Root MSE	LAI Mean		
0.99	0.41	0.01	2.42		
Source	DF	Type III SS	Mean square	F value	Pr > F
Treatment	6	2.13	0.36	3551	<.0001

Seed filling

Source	DF	Sum of squares	Mean square	F value	Pr > F
Model	6	4.73	0.79	1055	<.0001
Error	14	0.01	0.00		
Corrected total	20	4.74			
R-square	Coeff var	Root MSE	LAI Mean		
1.00	0.50	0.03	5.46		
Source	DF	Type III SS	Mean square	F value	Pr > F
Treatment	6	4.73	0.79	1055	<.0001

Maturity

Source	DF	Sum of squares	Mean square	F value	Pr > F
Model	6	9.08	1.52	4202	<.0001
Error	15	0.01	0.00		
Corrected total	21	9.08			
R-square	Coeff var	Root MSE	LAI Mean		
1.00	1.72	0.02	1.10		
Source	DF	Type III SS	Mean square	F value	Pr > F
Treatment	6	9.08	1.52	4202	<.0001

Appendix G. Analysis of variance (ANOVA) fractions of PAR at different stages of the growth of the crop in (a) 2011 and (b) 2012

(a)

VE-V2

Source	DF	Sum of squares	Mean square	F value	Pr > F
Model	6	0.01	0.00	21.7	<.0001
Error	14	0.00	0.00		
Corrected total	20	0.01			
R-square	Coeff var	Root MSE	fIPAR Mean		
0.90	3.26	0.01	0.25		
Source	DF	Type III SS	Mean square	F value	Pr > F
Treatment	6	0.01	0.00	21.7	<.0001

Flowering

Source	DF	Sum of squares	Mean square	F value	Pr > F
Model	6	0.03	0.00	68.4	<.0001
Error	14	0.00	0.00		
Corrected total	20	0.03			
R-square	Coeff var	Root MSE	fIPAR Mean		
0.97	2.36	0.01	0.33		
Source	DF	Type III SS	Mean square	F value	Pr > F
Treatment	6	0.03	0.00	68.4	<.0001

Pod initiation

Source	DF	Sum of squares	Mean square	F value	Pr > F
Model	6	0.04	0.01	67.21	<.0001
Error	14	0.00	0.00		
Corrected total	20	0.04			
R-square	Coeff var	Root MSE	fIPAR Mean		
0.97	1.46	0.01	0.69		
Source	DF	Type III SS	Mean square	F value	Pr > F
Treatment	6	0.04	0.01	67.21	<.0001

Seed filling

Source	DF	Sum of squares	Mean square	F value	Pr > F
Model	6	0.05	0.01	126	<.0001
Error	14	0.00	0.00		
Corrected total	20	0.05			
R-square	Coeff var	Root MSE	fIPAR Mean		
0.98	0.95	0.01	0.83		
Source	DF	Type III SS	Mean square	F value	Pr > F
Treatment	6	0.05	0.01	126	<.0001

Maturity

Source	DF	Sum of squares	Mean square	F value	Pr > F
Model	6	0.23	0.04	15.6	<.0001
Error	14	0.03	0.00		
Corrected total	20	0.27			
R-square	Coeff var	Root MSE	fIPAR Mean		
0.87	9.28	0.05	0.54		
Source	DF	Type III SS	Mean square	F value	Pr > F
Treatment	6	0.23	0.04	15.6	<.0001

(b)

VE-V2

Source	DF	Sum of squares	Mean square	F value	Pr > F
Model	6	0.02	0.00	128	<.0001
Error	14	0.00	0.00		
Corrected total	20	0.02			
R-square	Coeff var	Root MSE	fIPAR Mean		
0.98	3.16	0.01	0.17		
Source	DF	Type III SS	Mean square	F value	Pr > F
Treatment	6	0.02	0.00	128	<.0001

Flowering

Source	DF	Sum of squares	Mean square	F value	Pr > F
Model	6	0.05	0.01	112	<.0001
Error	14	0.00	0.00		
Corrected total	20	0.05			
R-square	Coeff var	Root MSE	fIPAR Mean		
0.98	2.05	0.01	0.41		
Source	DF	Type III SS	Mean square	F value	Pr > F
Treatment	6	0.05	0.01	112	<.0001

Pod initiation

Source	DF	Sum of squares	Mean square	F value	Pr > F
Model	6	0.04	0.01	1.54	0.24
Error	14	0.06	0.00		
Corrected total	21	0.10			
R-square	Coeff var	Root MSE	fIPAR Mean		
0.40	8.39	0.06	0.75		
Source	DF	Type III SS	Mean square	F value	Pr > F
Treatment	6	0.04	0.01	1.54	0.24

Seed filling

Source	DF	Sum of squares	Mean square	F value	Pr > F
Model	6	0.02	0.00	33.59	<.0001
Error	14	0.00	0.00		
Corrected total	20	0.03			
R-square	Coeff var	Root MSE	fIPAR Mean		
0.94	1.20	0.01	0.91		
Source	DF	Type III SS	Mean square	F value	Pr > F
Treatment	6	0.02	0.00	33.59	<.0001

Maturity

Source	DF	Sum of squares	Mean square	F value	Pr > F
Model	6	0.48	0.01	538	<.0001
Error	15	0.00	0.00		
Corrected total	21	0.48			
R-square	Coeff var	Root MSE	fIPAR Mean		
1.00	3.50	0.01	0.35		
Source	DF	Type III SS	Mean square	F value	Pr > F
Treatment	6	0.48	0.01	538	<.0001

Appendix H. Partitioning of evapotranspiration into transpiration and evaporation for water conservation in 2012

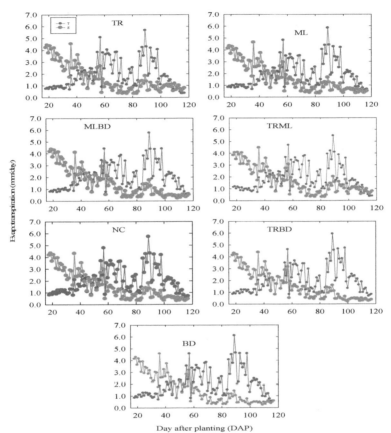

Figure F.1. Partitioning of evapotranspiration into transpiration and evaporation for water conservation in 2012

Appendix I. Analysis of variance of the yield of the crop in (a) 2011 and (b) 2012

2011

Source	DF	Sum of squares	Mean square	F value	Pr > F
Model	6	5.65	0.95	3.09	0.03
Error	21	6.40	0.31		
Corrected total	27	12.1			
R-square	Coeff var	Root MSE	Seed yield mean		
0.47	25.0	0.55	2.21		
Source	DF	Type III SS	Mean square	F value	Pr > F
Treatment	6	5.65	0.95	3.09	0.03

2012

Source	DF	Sum of squares	Mean square	F value	Pr > F
Model	6	8.38	1.40	4.12	0.01
Error	21	7.12			
Corrected total	27	15.5			
R-square	Coeff var	Root MSE	Seed yield mean		
0.54	24.4	0.58	2.38		
Source	DF	Type III SS	Mean square	F value	Pr > F
Treatment	6	8.38	1.40	4.12	0.01

Appendix 1. Analysis of variance of the yield of the crop in (a) 2011 and (b) 2012

Appendix J. ANOVA of water productivity of the crop in (a) 2011 and (b) 2012

(a)   2011

Source	DF	Sum of squares	Mean square	F value	Pr > F
Model	6	33.7	5.62	2.00	0.11
Error	21	59.1	2.81		
Corrected total	27	92.8			
R-square	Coeff var	Root MSE	WP Mean		
0.36	27.0	1.68	6.20		
Source	DF	Type III SS	Mean square	F value	Pr > F
Treatment	6	33.7	5.62	2.00	0.11

(b)   2012

Source	DF	Sum of squares	Mean square	F value	Pr > F
Model	6	23.2	3.87	3.64	0.01
Error	21	22.3	1.06		
Corrected total	27	45.5			
R-square	Coeff var	Root MSE	WP Mean		
0.51	24.3	1.03	4.25		
Source	DF	Type III SS	Mean square	F value	Pr > F
Treatment	6	23.2	3.87	3.64	0.01

Appendix K. ANOVA of harvest indices of the crop in 2011 and 2012 under rainfed conditions

2011

Source	DF	Sum of squares	Mean square	F value	Pr > F
Model	6	591	98.5	5.58	0.00
Error	21	371	17.7		
Corrected total	27	962			
R-square	Coeff var	Root MSE	HI Mean		
0.61	8.09	4.20	51.9		
Source	DF	Type III SS	Mean square	F value	Pr > F
Treatment	6	591	98.5	5.58	0.00

2012

Source	DF	Sum of squares	Mean square	F value	Pr > F
Model	6	105	17.5	0.68	0.67
Error	21	543	25.9		
Corrected total	27	648			
R-square	Coeff var	Root MSE	HI Mean		
0.16	9.08	5.09	56.0		
Source	DF	Type III SS	Mean square	F value	Pr > F
Treatment	6	105	17.5	0.68	0.67

Appendix L. Cost analysis for the six water conservation and the conventional practices under rainfed conditions

Operations	Treatment and cost (US$ ha^{-1})						
	TR	BD	ML	MLBD	TRML	TRBD	NC
Land Clearing	311	311	311	311	311	311	311
Ridging (manual)	391	260	0	391	391	391	0
Weeding	372	372	372	372	372	372	372
Chemicals (Insecticide)	75	75	75	75	75	75	75
Bund/ridge repair	271	174	-	271	271	271	-
Sub-Total	*1418*	*1191*	*757*	*1418*	*1418*	*1418*	*757*
Seed	17.4	17.4	17.4	17.4	17.4	17.4	17.4
Planting (manual)	123	123	123	123	123	123	123
Sub-Total	*141*	*141*	*141*	*141*	*141*	*141*	*141*
Harvesting(manual)	248	248	248	248	248	248	248
Threshing (manual)	379	379	379	379	379	379	379
Transportation	93.8	93.8	93.8	93.8	93.8	93.8	93.8
Sub-Total	*720*	*720*	*720*	*720*	*720*	*720*	*720*
Total cost US$	2280	2052	1620	2280	2280	2280	1620

Appendix M. ANOVA of LAIs in the dry season

2013 irrigation season

49 DAP

Source	DF	Sum of squares	Mean square	F value	Pr > F
Model	4	2.35	0.59	3.15	0.06
Error	10	1.87	0.19		
Corrected total	14	4.22			
R-square	Coeff var	Root MSE	LAI Mean		
0.55	13.8	0.43	0.43		
Source	DF	Type III SS	Mean square	F value	Pr > F
Treatment	4	2.36	0.59	3.15	0.06

63 DAP

Source	DF	Sum of squares	Mean square	F value	Pr > F
Model	4	1.55	0.39	3.33	0.06
Error	10	1.17	0.12		
Corrected total	14	2.72			
R-square	Coeff var	Root MSE	LAI Mean		
0.57	7.02	0.34	4.87		
Source	DF	Type III SS	Mean square	F value	Pr > F
Treatment	4	1.55	0.39	3.33	0.06

86 DAP

Source	DF	Sum of squares	Mean square	F value	Pr > F
Model	4	21.56	5.39	27.6	0.00
Error	10	1.95	0.20		
Corrected total	14	23.5			
R-square	Coeff var	Root MSE	LAI Mean		
0.92	9.13	0.44	4.84		
Source	DF	Type III SS	Mean square	F value	Pr > F
Treatment	4	21.6	5.39	27.6	0.00

109 DAP

Source	DF	Sum of squares	Mean square	F value	Pr > F
Model	4	6.36	1.59	49.0	0.00
Error	10	0.32	0.03		
Corrected total	14	6.68			
R-square	Coeff var	Root MSE	LAI Mean		
0.95	16.7	0.18	1.08		
Source	DF	Type III SS	Mean square	F value	Pr > F
Treatment	4	6.35	1.59	49.0	0.00

2013/2014 irrigation season

49 DAP

Source	DF	Sum of squares	Mean square	F value	Pr > F
Model	4	1.40	0.35	4.42	0.07
Error	5	0.39	0.08		
Corrected total	9	1.79			
R-square	Coeff var	Root MSE	LAI Mean		
0.78	9.61	0.28	2.92		
Source	DF	Type III SS	Mean square	F value	Pr > F
Treatment	4	1.40	0.35	4.42	0.07

56 DAP

Source	DF	Sum of squares	Mean square	F value	Pr > F
Model	4	0.31	0.08	2.80	0.14
Error	5	0.14	0.08		
Corrected total	9	0.45			
R-square	Coeff var	Root MSE	LAI Mean		
0.69	7.04	0.17	2.36		
Source	DF	Type III SS	Mean square	F value	Pr > F
Treatment	4	0.31	0.08	2.80	0.14

98 DAP

Source	DF	Sum of squares	Mean square	F value	Pr > F
Model	4	0.15	0.04	1.65	0.30
Error	5	0.11	0.02		
Corrected total	9	0.25			
R-square	Coeff var	Root MSE	LAI Mean		
0.57	13.3	0.15	1.10		
Source	DF	Type III SS	Mean square	F value	Pr > F
Treatment	4	0.15	0.04	1.65	0.30

109 DAP

Source	DF	Sum of squares	Mean square	F value	Pr > F
Model	4	0.33	0.08	592	<0.00
Error	5	0.00	0.00		
Corrected total	9	0.33			
R-square	Coeff var	Root MSE	LAI Mean		
0.99	1.59	0.011	0.74		
Source	DF	Type III SS	Mean square	F value	Pr > F
Treatment	4	0.33	0.08	592	<0.00

Appendix N. ANOVA of dry matter in the dry season

2013 irrigation season (Note: the analysis was done in t ha^{-1})

49 DAP

Source	DF	Sum of squares	Mean square	F value	Pr > F
Model	4	0.60	0.15	2.00	0.17
Error	10	0.74	0.07		
Corrected total	14	1.34			
R-square	Coeff var	Root MSE	DAB Mean		
0.44	27.3	0.27	1.00		
Source	DF	Type III SS	Mean square	F value	Pr > F
Treatment	4	0.60	0.15	2.00	0.17

63 DAP

Source	DF	Sum of squares	Mean square	F value	Pr > F
Model	4	1.15	0.29	3.92	0.04
Error	10	0.73	0.07		
Corrected total	14	1.88			
R-square	Coeff var	Root MSE	DAB Mean		
0.61	13.5	0.27	2.00		
Source	DF	Type III SS	Mean square	F value	Pr > F
Treatment	4	1.15	0.29	3.92	0.04

84 DAP

Source	DF	Sum of squares	Mean square	F value	Pr > F
Model	4				
Error	5				
Corrected total	9				
R-square	Coeff var	Root MSE	DAB Mean		
Source	DF	Type III SS	Mean square	F value	Pr > F
Treatment					

109 DAP

Source	DF	Sum of squares	Mean square	F value	Pr > F
Model	4	6.04	1.52	3.34	0.06
Error	5	4.53	0.46		
Corrected total	9	10.6			
R-square	Coeff var	Root MSE	DAB Mean		
0.57	14.6	0.68	4.61		
Source	DF	Type III SS	Mean square	F value	Pr > F
Treatment	4	6.04	1.51	3.34	0.06

2013/2014 irrigation season 49 DAP

49 DAP

Source	DF	Sum of squares	Mean square	F value	Pr > F
Model	4	607	152	0.52	0.73
Error	5	1456	292		
Corrected total	9	2063			
R-square	Coeff var	Root MSE	DAB Mean		
0.29	24.6	17.1	69.3		
Source	DF	Type III SS	Mean square	F value	Pr > F
Treatment	4	607	152	0.52	0.73

63 DAP

Source	DF	Sum of squares	Mean square	F value	Pr > F
Model	4	3207	802	2.66	0.16
Error	5	1508	302		
Corrected total	9	4714			
R-square	Coeff var	Root MSE	DAB Mean		
0.68	20.2	17.4	86.2		
Source	DF	Type III SS	Mean square	F value	Pr > F
Treatment	4	3206	802	2.66	0.16

84 DAP

Source	DF	Sum of squares	Mean square	F value	Pr > F
Model	4	8952	2238	1.17	0.42
Error	5	9528	1906		
Corrected total	9	18479			
R-square	Coeff var	Root MSE	DAB Mean		
0.48	25.8	43.7	169		
Source	DF	Type III SS	Mean square	F value	Pr > F
Treatment	4	8952	2238	1.17	0.42

109 DAP

Source	DF	Sum of squares	Mean square	F value	Pr > F
Model	4	961	240	126	< 0.00
Error	5	9.50	0.41.9		
Corrected total	9	970			
R-square	Coeff var	Root MSE	DAB Mean		
0.99	1.19	1.38	116		
Source	DF	Type III SS	Mean square	F value	Pr > F
Treatment	4	961	240	126	< 0.00

Appendix O. ANOVA of seed yields in the dry season

2013 irrigation season

Source	DF	Sum of squares	Mean square	F value	Pr > F
Model	4	3.05	0.76	3.74	0.04
Error	10	2.03	0.20		
Corrected total	14	5.07			
R-square	Coeff var	Root MSE	Seed yield mean		
0.60	18.2	0.45	2.48		
Source	DF	Type III SS	Mean square	F value	Pr > F
Treatment	4	3.05	0.76	3.74	0.04

2013/2014 irrigation season

Source	DF	Sum of squares	Mean square	F value	Pr > F
Model	4	1.05	0.26	3.70	0.04
Error	10	0.71	0.07		
Corrected total	14	1.76			
R-square	Coeff var	Root MSE	Seed yield mean		
0.60	23.2	0.27	1.15		
Source	DF	Type III SS	Mean square	F value	Pr > F
Treatment	4	1.05	0.26	3.70	0.04

Appendix P. Analysis of the cost of production under irrigation conditions

Scenario (a). Situation whereby the cost of water is included in the total cost of production

Operations/Costs	Treatment/cost per ha (US$)				
Fixed Costs	T_{1111}	T_{0111}	T_{1011}	T_{1101}	T_{1110}
Water tank + plumbing	32	32	32	32	32
Drip lines + accessories (with shipping and custom duties)	4545	4545	4545	4545	4545
Pumping machine (5.5 HP + PVC hose (100 m)	32.7	32.7	32.7	32.7	32.7
Sub total	4609.7	4609.7	4609.7	4609.7	4609.7
Variable cost					
Plough	41.4	41.4	41.4	41.4	41.4
Chemical (insecticide)	19.4	19.4	19.4	19.4	19.4
Weeding (manual)	38.9	38.9	38.9	38.9	38.9
Sub total	*99.7*	*99.7*	*99.7*	*99.7*	*99.7*
Water	7320	6870	6870	5940	6870
Seed	24.0	24.0	24.0	24.0	24.0
Planting (manual)	64.8	64.8	64.8	64.8	64.8
Harvesting (manual)	77.7	77.7	77.7	77.7	77.7
Threshing (manual)	129.5	129.5	129.5	129.5	129.5
Transportation	19.4	19.4	19.4	19.4	19.4
Sub total	*315*	*315*	*315*	*315*	*315*
Total US$	12300	11900	11900	11000	11900

Scenario (b). Situation whereby water is provided free of charge by an agency of government or pumped from a stream without cost

Operations/costs	Treatment/cost per ha (US$)				
Fixed cost	T_{1111}	T_{0111}	T_{1011}	T_{1101}	T_{1110}
Water tank + plumbing	32	32	32	32	32
Drip lines + accessories (with shipping and custom duties)	4545	4545	4545	4545	4545
Pumping machine (5.5 HP)+ PVC hose (100 m)	32.7	32.7	32.7	32.7	32.7
Sub total	4609.7	4609.7	4609.7	4609.7	4609.7
Variable cost					
Plough and harrow	41.4	41.4	41.4	41.4	41.4
Chemical (insecticide)	19.4	19.4	19.4	19.4	19.4
Weeding (manual)	38.9	38.9	38.9	38.9	38.9
Fuelling of pumping machine (petrol)	987	900	900	675	900
Sub total	1086	999.7	999.7	774.7	999.7
Seed	24.0	24.0	24.0	24.0	24.0
Planting (manual)	64.8	64.8	64.8	64.8	64.8
Harvesting (manual)	77.7	77.7	77.7	77.7	77.7
Threshing (manual)	129.5	129.5	129.5	129.5	129.5
Transportation	19.4	19.4	19.4	19.4	19.4
Sub total	*315*	*315*	*315*	*315*	*315*
Total US$	6010.7	5924.5	5924.5	5699.4	5924.5

Appendix Q. Samenvatting

De wereld bevolking neemt sneller toe dan ooit in de geschiedenis. Om te voldoen aan de voedselbehoefte, met name in Sub-Sahara Afrika, is er een dringende noodzaak om de voedselproductie te vergroten,. In het stroomgebied van de Ogun-Osun rivier, Nigeria, wordt meer dan 95% van de gewassen onder regenafhankelijke condities geteeld. Fluctuatie in regenval in het stroomgebied als gevolg van klimaatverandering is een belangrijke uitdaging in de afgelopen tijd. Productiviteit van de grond kan sterk worden verbeterd door het toepassen van betaalbare water besparende maatregelen door de boeren die in het stroomgebied de gewassen telen. Eveneens zouden waterbesparende maatregelen kunnen worden toegepast door het gebruik van druppelbevloeiing en het toedienen van water tijdens kritieke fasen van de groei van gewassen. Vruchtbaarheid van de bodem dient te worden gehandhaafd door het telen van gewassen die op natuurlijke wijze voedingsstoffen aan de bodem toevoegen. Dergelijke maatregelen kunnen een belangrijke bijdragen leveren om te komen tot duurzaam gebruik van land en water in het stroomgebied van de Ogun-Osun rivier.

De onbepaalde variëteit van sojabonen TGX 1448 2E is geteeld op de boerderijen voor onderwijs en onderzoek van de Obafemi Awolowo Universiteit, Ile-Ife, Nigeria tijdens de natte seizoenen van mei tot september in 2011 en juni tot oktober in 2012. Ook is het gewas onder druppelbevloeiing gedurende twee seizoenen geteeld van februari tot mei 2013 en van november 2013 tot februari 2014. Het doel van het uitvoeren van de experimenten in de natte en droge seizoenen was om de opbrengst en hun componenten te vergelijken en om de opbrengsten van het gewas te beoordelen in termen van het watergebruik en productiviteit. Het experimentele veld tijdens het droge seizoen lag op ongeveer 1 km van het veld tijdens het natte seizoen. Dit was in verband met de dichtbij zijnde de bron voor het water. Tijdens de experimenten in de vier seizoenen zijn de belangrijkste biometrische gegevens van het gewas van zaaien tot fysiologische rijpheid gemeten. De gewas cyclus tijdens de experimenten in de natte tijd duurde 117 en 119 dagen in 2011 en 2012, terwijl in de droge tijd deze in het eerste seizoen 112 dagen duurde en 105 dagen in het tweede seizoen. De lengte van de gewas cycli in de vier seizoenen verschilde een beetje. Dit wordt toegeschreven aan omgevingsfactoren, zoals weersomstandigheden, beschikbaarheid van voedingsstoffen in de bodem en de periode van de teelt. Tijdens het natte seizoen, werden zes waterbesparende behandelingen toegepast namelijk richels, afdekken met plantmaterialen, grond ruggen, richels en grondruggen, richels met afdekken, afdekken en grondruggen en direct zaaien zonder waterbesparende maatregelen (conventionele praktijk), dit was de controle behandeling. De behandelingen werden toegepast in een veld van 31 bij 52 m (1.612 m^2) met een willekeurige verdeling van de blokken in viervoud en standaard agronomische maatregelen. Voor het bepalen van de gewasverdamping is tijdens zowel de natte als de droge seizoenen de aanpak van de waterbalans voor de bodem toegepast. De gewasverdamping is onderverdeeld in de productieve gewasverdamping en de niet-productieve verdamping vanuit de bodem.

Seizoensgemiddelde bedekking extinctiecoëfficiënten van het gewas waren respectievelijk 0,46 en 0,51 in de natte seizoenen van 2011 en 2012, terwijl ze in de droge seizoenen van 2013 en 2013/2014 respectievelijk 0,43 en 0,49 waren. De plant hoogte varieerde in 2011 van 51,3 cm voor grondruggen tot 67,8 cm voor de conventionele praktijk, terwijl de hoogte in 2012 varieerde van 60,3 cm voor richels met grondruggen tot 80,3 cm voor afdekken met grondruggen. De minimale fractie van onderschepte fotosynthetisch actieve straling was 0,13 voor plantvorming bij richels en grondruggen, terwijl de piek fractie van onderschepte fotosynthetisch actieve straling bij zaadvorming tijdens het natte seizoen 0,97 voor grondruggen was. Overeenkomstig waren de minimale en maximale bladoppervlak indicatoren 0,13 m² m⁻² voor richels en grondruggen tijdens de plantvorming in 2011 en 6,61 m² m⁻² voor grondruggen tijdens de zaadvorming in 2012. Bij toepassing van een exponentieel model waren er sterke en significante correlaties tussen de fractie van onderschepte fotosynthetisch actieve straling en de bladoppervlakte indicatoren in 2011 ($0,70 \leq r^2 \leq 0,99$) en in 2012 ($0,93 \leq r^2 \geq 0.99$). Seizoensgebonden regenval was respectievelijk 539 en 761 mm in 2011 en 2012. Seizoensgebonden berging van water in de bodem varieerde in 2011 van 407 mm voor de conventionele praktijk tot 476 mm voor richels en afdekken, terwijl dit in 2012 varieerde van 543 mm voor richels tot 578 mm voor richels en grondruggen.

Straling efficiëntie werd bepaald door het voor alle behandelingen plotten van de met intervallen van zeven dagen gemeten droge bovengrondse biomassa tegen de dagelijkse fotosynthetisch effectieve straling van de zonnestraling en de onmiddellijke fotosynthetisch actieve straling gemeten in de buurt van de zonnemiddag. Voor de fotosynthetisch effectieve straling afkomstig van de zonnestraling, varieerde de straling efficiëntie van het gewas van 1,18 g MJ⁻¹ voor richels tot 1,98 g MJ⁻¹ van de onderschepte fotosynthetisch actieve straling voor richels en grondruggen in 2011, terwijl deze in 2012 varieerde van 1,45 g MJ⁻¹ voor richels tot 1,92 g MJ⁻¹ voor afdekken. Er is tussen de twee seizoenen geen significant verschil gevonden in de gemiddelde seizoensgebonden straling efficiëntie. Bij de momentane meting van de fotosynthetisch effectieve straling varieerde de straling efficiëntie in 2011 van 0,80 g MJ⁻¹ van onderschepte fotosynthetisch actieve straling voor richels tot 1,65 g MJ⁻¹ voor richels en grondruggen, terwijl deze in 2012 varieerde van 0,94 g MJ⁻¹ voor richels tot 1,24 g MJ⁻¹ voor grondruggen. De twee benaderingen gaven relatief gelijkaardige waarden voor straling efficiëntie. Voor de twee seizoenen zijn positieve correlatie coëfficiënten ($0,50 \leq r^2 \leq 0,89$) gevonden voor de behandelingen tussen de met een licht model gesimuleerde droge bovengrondse biomassa en de in het veld gemeten waarden.

Het seizoensgebonden watergebruik door het gewas varieerde in 2011 van 311 mm voor afdekken en grondruggen tot 406 mm voor richels en grondruggen, terwijl het in 2012 varieerde van 533 mm voor afgedekte percelen tot 589 mm voor grondruggen. Seizoensgebonden gewasverdamping varieerde in 2011 van 190 mm voor richels en afdekken tot 204 mm voor grondruggen, terwijl het in 2012 varieerde van 164 mm voor richels en afdekken tot 195 mm voor afgedekte percelen. Seizoensgebonden verdamping was in 2012 hoger, variërend van 338 mm voor afgedekte percelen tot 408 mm voor grondruggen, terwijl het in 2011 varieerde van 311 mm voor afdekken en grondruggen tot 406 mm voor richels en grondruggen. Berging van water in de bodem en seizoensgebonden watergebruik door het gewas zijn gerelateerd. Overeenkomstig waren over de twee seizoenen het seizoensgebonden watergebruik door het gewas, de onderschepte fotosynthetisch actieve straling en de straling efficiëntie voor het gewas sterk gerelateerd.

De verhandelbare zaad opbrengst varieerde in 2011 van 1.68±0.50 t ha⁻¹ voor richels tot 2,95±0,30 t ha⁻¹ voor richels en grondruggen, terwijl de opbrengst in 2012 varieerde van 1.64±0.50 t ha⁻¹ voor de conventionele praktijk tot 3,25±0,52 t ha⁻¹ voor afdekken en grondruggen. In 2011 was de zaadopbrengst voor richels en grondruggen 15,6, 15,9, 25,4, 28,5, 43,1 en 47,1% hoger dan de zaadopbrengst voor respectievelijk afdekken en grondruggen, grondruggen, afdekken, richels en afdekken, richels en de conventionele praktijk. In 2012 was de zaadopbrengst voor afdekken en grondruggen 7,4, 21,8, 32,0, 32,3, 43,7 en 49,5% hoger dan de zaadopbrengst voor respectievelijk grondruggen, richels, afdekken, richels en afdekken, richels en grondruggen en direct zaaien. De gemiddelde seizoengebonden zaadopbrengst van het gewas was significant gerelateerd aan de totale onderschepte fotosynthetisch actieve straling, maar niet aan de straling efficiëntie. Oogst indicatoren varieerden in 2011 van 47,4±4,5% voor richels tot 57,6±1,1% voor richels en grondruggen en in 2012 van 53,1±3,0% voor grondruggen tot 58,1±2,3% voor richels De hoogste oogst indicatoren werden in respectievelijk 2011 en 2012 verkregen bij richels en grondruggen en bij richels. De oogstindex was niet significant gerelateerd aan zowel de onderschepte fotosynthetisch actieve straling en de straling efficiëntie van het gewas.

Voor alle behandelingen waren de gemiddelde efficiëntie van seizoengebonden gewasverdamping, dat is de verhouding van de droge bovengrondse biomassa bij het oogsten tot de seizoensgebonden gewasverdamping, 7,0 kg ha^{-1} mm^{-1} in 2011 en 14,9 kg ha^{-1} mm^{-1} in 2012. Efficiëntie van gewasverdamping was sterk gerelateerd aan de onderschepte fotosynthetisch actieve straling, maar niet aan de straling efficiëntie onder veldomstandigheden in het natte seizoenen. De hoogste water productiviteit voor zaad was 7.99 kg ha^{-1} mm^{-1} in 2011 en 5,76 kg ha^{-1} mm^{-1} in 2012 voor afdekken en grondruggen. De water productiviteit voor zaad was sterk en significant gerelateerd aan de onderschepte fotosynthetisch actieve straling. Het was echter niet significant gerelateerd aan de straling efficiëntie. Deze bevindingen zullen informatie opleveren voor bouwers van modellen betreffende gewasopbrengst bij het simuleren van opbrengsten van sojabonen bij water besparende maatregelen.

De aanleg van richels en grondruggen speciaal voor richels, afdekken en grondruggen en richels en grondruggen verhoogde de gemiddelde seizoensgebonden productiekosten met 28,9% in vergelijking met afdekken en de conventionele praktijk en met 10,1% ten opzichte van grondruggen. Daarnaast was de economische water productiviteit 3.90 US$ ha^{-1} mm^{-1} voor afdekken en grondruggen, terwijl voor grondruggen en de conventionele praktijk, deze respectievelijk 3,30 en 2,27 US$ ha^{-1} mm^{-1} was.

Vanwege de toename van de vraag naar voedsel, is er de noodzaak om meer gewas per druppel water onder regen afhankelijke omstandigheden te produceren en om het waterbeheer voor de landbouw te realiseren op stroomgebied niveau. De belangrijkste prioriteit in het studiegebied was om de zaad opbrengsten, de water en economische productiviteit en de financiële voordelen aan het einde van een groeiseizoen te verhogen. De resultaten tonen aan dat het toepassen van afdekken en grondruggen leidde tot de gemiddelde maximale efficiëntie van gewasverdamping, zaadopbrengst, water en economische productiviteit en een opbrengst van 1.630 US$ per ha. Bij het vergelijken van de gemiddelde seizoensgebonden efficiëntie va gewasverdamping, water gebruik door het gewas, opbrengst, water productiviteit en de productie kosten voor de zes waterbesparende maatregelen met die van de gebruikelijke praktijk in de twee natte seizoenen, had afdekken en grondruggen de maximale gemiddelde zaadopbrengst, water en economische productiviteit. Afdekken en grondruggen wordt hierbij aanbevolen voor de teelt van het gewas in het studiegebied. Andere waterbesparende maatregelen, zoals grondruggen, presteerden ook naar tevredenheid in termen van zaadopbrengst en water productiviteit, hoewel met een lichte daling van de opbrengst. Het gebruik van deze waterbesparende maatregelen zal niet alleen de opbrengsten van het gewas verhogen, maar ook de uitputting van water in de bodem, wat zou kunnen leiden tot het initiëren of aantasten van de bodem in het studiegebied, tot het absolute minimum beperken. Daarom kan de duurzaamheid van land en water in het stroomgebied van de Ogun-Osun rivier worden gewaarborgd. Deze bevindingen tonen aan dat land en water productiviteit van sojabonen onder regenafhankelijke omstandigheden aanzienlijk verbeterd kan worden met waterbesparende maatregelen onder de huidige variaties in regen en de concurrentie om natuurlijke rijkdommen tussen landbouw en stedelijk grondgebruik in het stroomgebied van de Ogun-Osun rivier.

Ook voor twee irrigatie seizoenen van februari tot mei 2013 en november 2013 tot februari 2014 zijn veldproeven uitgevoerd. Het gewas werd geplant in een willekeurig compleet ontwerp van de blokken met drie herhalingen en in lijn druppelirrigatie werd gebruikt om water naar de gewassen te brengen. Vijf behandelingen zijn geselecteerd, te weten: (i) volledige irrigatie, het overslaan van irrigatie om de andere week tijdens: (ii) de bloei; (iii) vorming van peulen; (iv) zaadvorming; (v) afrijpen. Biometrische gegevens betreffende het aantal blaadjes, planthoogte, bladoppervlak indicatoren en droge bovengrondse biomassa zijn in de twee irrigatie seizoenen elke week bepaald en vastgelegd van zaaien tot afrijpen. Bodem vochtgehalten zijn voorafgaand aan de irrigatie bij de wortelzone van de planten bepaald om de netto irrigatiewater behoefte in elk groeistadium te bepalen. Oogst indicatoren zijn voor elke behandeling bepaald. Aantal peulen per plant, aantal zaden per peul en de opbrengst bij elke behandeling zijn na de fysiologische rijpheid in elk seizoen bepaald. Regressievergelijkingen zijn gegenereerd voor: (i) de opbrengst; (ii) aantal peulen per plant; (iii) aantal zaden per peul; (iv) aantal bladeren; (v) seizoensgebonden gewasverdamping en bladoppervlak indicatoren. Op overeenkomstige wijze zijn regressievergelijkingen gegenereerd voor: (i) planthoogten; (ii) seizoensgebonden gewasverdamping; (iii) aantal peulen per plant; (iv) aantal zaden per peul; (v) droge bovengrondse biomassa. Lineaire regressies zijn ook bepaald voor de opbrengst, droge bovengrondse biomassa en seizoensgebonden watergebruik door het gewas. De gewas reactie factor is bepaald. Water productiviteit en irrigatie water productiviteit zijn voor elke behandeling berekend en vergeleken. Lineaire modellen zijn bepaald betreffende de water productiviteit, irrigatie water productiviteit en de oogst index.

De bijdrage van de regen aan het watergebruik door het gewas was respectievelijk 262 en 50 mm voor de 2013 en 2013/2014 irrigatie seizoenen. De maximum bladoppervlakte indicator in het 2013 irrigatieseizoen was 7.10 m^2 m^{-2} voor volledige irrigatie bij zaadvorming, terwijl het in het 2013/2014 irrigatieseizoen 3,44 m^2 m^{-2} was bij volledige irrigatie gedurende de bloei. De droge bovengrondse biomassa na afrijpen varieerde van 359 g m^{-2}, wanneer irrigatie bij het begin van afrijpen om de week werd overgeslagen tot 578 g m^{-2} voor volledige irrigatie. De zaad opbrengsten varieerden van 1,81 t ha^{-1} wanneer de irrigatie tijdens zaadvorming om de week werd overgeslagen tot 3,11 t ha^{-1} voor volledige irrigatie. Gemiddelde seizoensgebonden zaadopbrengst voor volledige irrigatie was respectievelijk 18,8, 21,8, 24,4 en 47,9% hoger dan de opbrengsten voor behandelingen waar irrigatie om de week werd overgeslagen tijdens de bloei, vorming van peulen, aanvang van afrijpen en zaadvorming. Seizoensgebonden gewasverdamping varieerde van 217 mm wanneer irrigatie tijdens zaadvorming om de week werd overgeslagen tot 409 mm voor volledige irrigatie in het 2013 irrigatie seizoen, terwijl in het 2013/2014 irrigatieseizoen het varieerde van 28 mm voor de behandeling waar irrigatie om de week tijdens zaadvorming werd overgeslagen tot 223 mm voor volledige irrigatie. Seizoensgebonden watergebruik door het gewas varieerde van 463 mm wanneer de irrigatie tijdens de bloei om de week werd overgeslagen tot 523 mm voor volledige irrigatie in het 2013 irrigatie seizoen, terwijl het in het 2013/2014 irrigatieseizoen varieerde van 364 mm wanneer de irrigatie om de week gedurende zaadvorming werd overgeslagen tot 507 mm bij volledige irrigatie. Oogst indicatoren varieerden van 56,0% wanneer irrigatie werd overgeslagen tijdens zaadvorming tot 65,9% wanneer irrigatie werd overgeslagen tijdens de bloei in het 2013 irrigatie seizoen, terwijl het in het 2013/2014 irrigatie seizoen varieerde van 43,2% wanneer irrigatie tijdens zaadvorming werd overgeslagen tot 63,9% bij volledige irrigatie. Water productiviteit voor de productie van zaaizaad varieerde van 3,89 kg ha mm^{-1} wanneer irrigatie bij zaadvorming werd overgeslagen tot

5,95 kg ha^{-1} mm^{-1} voor volledige irrigatie in het 2013 irrigatie seizoen, terwijl het in het 2013/2014 irrigatie seizoen varieerde van 1,93 kg ha mm^{-1} wanneer irrigatie tijdens zaadvorming werd overgeslagen tot 3,00 kg ha^{-1} mm^{-1} bij volledige irrigatie. Irrigatie water productiviteit varieerde van 8.90 kg ha^{-1} mm^{-1} wanneer irrigatie tijdens zaadvorming werd overgeslagen tot 14,0 kg ha^{-1} mm^{-1} wanneer irrigatie tijdens de bloei werd overgeslagen in het 2013 irrigatie seizoen, terwijl het in het 2013/2014 seizoen varieerde van 2,24 kg ha^{-1} mm^{-1} wanneer irrigatie tijdens zaadvorming werd overgeslagen tot 3,32 kg ha^{-1} mm^{-1} bij volledige irrigatie. Bladoppervlak indicatoren en opbrengst, aantal bladeren, aantal peulen per plant, aantal zaden per peul en seizoensgebonden gewasverdamping waren significant gecorreleerd. Op overeenkomstige wijze waren de droge bovengrondse biomassa en seizoensgebonden gewasverdamping, aantal peulen per plant, aantal zaden per peul significant gecorreleerd. De gewasreactie factor (Ky), een maat voor de relatieve afname in zaadopbrengst door de relatieve vermindering van gewasverdamping, was 2,24. Dit geeft aan dat het aan het gewas opgelegde tekort aan irrigatiewater hoog was en dat de relatieve afname van de opbrengsten als gevolg van het tekort aan irrigatie water hoger was dan de relatieve daling van de gewasverdamping.

De resultaten tonen aan dat het overslaan van irrigatie tijdens elk groeistadium van het gewas leidde tot reductie van bladoppervlak indicatoren, droge bovengrondse biomassa en seizoengebonden watergebruik door het gewas. Onder irrigatie had significante effecten op zowel de droge stof en als de opbrengsten. Het effect van onder irrigatie was duidelijker op zaad opbrengsten dan op droge stof. De ernst van de gevolgen van onder irrigatie was afhankelijk van het groeistadium en de duur. Onder irrigatie verminderde aanzienlijk de droge stof bij de bloei en vorming van peulen. Onder irrigatie had echter geen invloed op de hoogte van de planten. Aantal zaden per plant tijdens de bloei en het begin van afrijpen werden aanzienlijk verminderd door onder irrigatie. Het aantal zaden per peul werd aanzienlijk verminderd wanneer de irrigatie gedurende de vorming van peulen werd overgeslagen. Zaad opbrengsten werden aanzienlijk verminderd wanneer de irrigatie tijdens zaadvorming werd overgeslagen. In het 2013 irrigatie seizoen was de water productiviteit wanneer irrigatie werd overgeslagen tijdens de bloei respectievelijk 2,3, 16,1, 23,5 en 36,1% hoger dan de water productiviteit bij volledige irrigatie, wanneer irrigatie werd overgeslagen tijdens vorming van peulen, begin van afrijpen en zaadvorming. In hetzelfde seizoen was irrigatie water productiviteit wanneer irrigatie werd overgeslagen tijdens de bloei respectievelijk 15, 20, 29,3 en 36,4% hoger dan bij volledige irrigatie, wanneer irrigatie werd overgeslagen tijdens de vorming van peulen, aanvang van afrijpen en de zaadvorming. In het 2013/2014 irrigatie seizoen was de water productiviteit echter bij volledige irrigatie respectievelijk 8,7, 16,3, 24,7 en 35,7% hoger dan de water productiviteit wanneer de irrigatie werd overgeslagen tijdens de vorming van peulen, aanvang van afrijpen, bloei en zaadvorming. Op overeenkomstige wijze was de irrigatie water productiviteit respectievelijk 7,2, 15,4, 24,1 en 32,5% hoger dan wanneer irrigatie werd overgeslagen tijdens de vorming van peulen, aanvang van afrijpen, bloei en zaadvorming. Bovendien was de irrigatie water productiviteit bij volledige irrigatie respectievelijk 24,1 en 32,5% hoger dan wanneer irrigatie werd overgeslagen tijdens de bloei en de zaadvorming. Stadium van de groei, de duur ervan, waterbehoeften en seizoensgebonden milieu omstandigheden beïnvloedden het seizoensgebonden watergebruik, de water productiviteit en irrigatiewater water productiviteit bij sojabonen. De maximale water productiviteit en irrigatie water productiviteit zijn in de twee seizoenen niet behaald bij volledige irrigatie. De maximale water productiviteit en irrigatie water productiviteit zijn behaald wanneer in het eerste seizoen de irrigatie alleen tijdens de bloei om de week werd overgeslagen, terwijl in het tweede seizoen volledige irrigatie de hoogste water en irrigatiewater productiviteit opleverde. Dit suggereert dat irrigatiewater productiviteit bij sojaboon kan worden verbeterd door het overslaan van irrigatie tijdens de bloei en de vorming van peulen.

In deze studie waren de productie kosten voor alle irrigatie scenario's hoog. Dit is te wijten aan de hoge kosten van water, die als het water moet worden gekocht tussen de 54-59% van de productie kosten waren en de kosten van druppelirrigatie apparatuur, die tussen 75,6-76,7% van de totale productie kosten lagen als het water zonder kosten beschikbaar zou zijn. Onder de heersende prijs en de economische omstandigheden na de oogst, levert het gebruik van in lijn druppelirrigatie geen economisch voordeel voor boeren op, die de belangrijkste telers van het gewas in Ile-Ife zijn. Economisch voordeel kan worden bereikt na lange perioden van gebruik en goed onderhoud van de irrigatie apparatuur en niet in rekening brengen van de vaste kosten.

Het op water gebaseerde gewasmodel AquaCrop is gekalibreerd en gevalideerd om de bedekkinggraad, droge bovengrondse biomassa, zaadopbrengst, gewasverdamping, gehalte van water in de bodem en water productiviteit van het gewas te voorspellen. De gesimuleerde en gemeten waarden komen goed overeen behalve voor de water productiviteit die in de validatie waarden te hoog werd voorspeld. Het AquaCrop model voorspelde de bedekkinggraad met fout statistieken van $0,93 \le e \le 0,98$ voor zowel volledige als onder irrigatie en de mate van overeenstemming $d = 0,99$ met $4,3 \le RMSE \le 5,9$ (de wortel van het kwadraat van de gemiddelde fout) voor volledige irrigatie, terwijl het voor onder irrigatie, $0,96 \le d \le 0,99$ met $5,3 \le RMSE \le 5,8$ was. Droge bovengrondse biomassa werd voorspeld met fout statistieken van $0,08 \le RMSE \le 0,14$ t ha^{-1} met $0,98 \le d \le 0,99$ voor volledige irrigatie, terwijl het voor het onder irrigatie $0,06 \le RMSE \le 1,09$ t ha^{-1} met $0,85 \le d \le 0,99$ was. 1 op elke 5 voorspellingen van de bovengrondse biomassa had meer dan 20% afwijking van de gemeten waarden.

De zaad opbrengsten zijn voorspeld met fout statistieken van $RMSE = 0,10$ t ha^{-1} en $d = 0,99$ en 1 op de 5 voorspellingen had meer dan 15% afwijking van de gemeten waarden. De fout statistiek in de voorspellingen voor seizoensgebonden water gebruik door het gewas voor zowel volledige als onder irrigatie behandelingen was in de twee seizoenen $15,4 \le RMSE \le 58.3$. Het AquaCrop model voorspelde te hoge waarden voor percolatie, ook in de validatie dataset. Deze waarnemingen suggereren dat de percolatie componenten van het model moeten worden aangepast om betere resultaten te krijgen. De resultaten van het model bij het voorspellen van bedekkinggraad, zaadopbrengst en andere grootheden in deze studie zijn prijzenswaardig en bevredigend.

Specifieke en duidelijke kenmerken, zoals het gebruik van bedekkinggraad in plaats van bladoppervlakte index, maken het model geschikt voor ontwikkelingslanden zoals Nigeria, waar onderzoekers vaak geen toegang kunnen hebben tot state-of-the-art apparatuur voor het meten van de bladoppervlakte index. Ook water productiviteit die is genormaliseerd voor de atmosferische behoefte en kooldioxide concentratie en de focus op water maakt het model geschikt voor diverse locaties. Door de jaren heen kan worden waargenomen dat geen enkel model universeel is in het vermogen om rekening te houden met alle verschillen in gewas variëteiten, milieu, weer en beheersomstandigheden. Andere variëteiten van sojabonen in Nigeria en andere agro klimatologische omstandigheden moeten worden getest en verfijnd in het model, om de nauwkeurigheid van het model vast te stellen. In het algemeen kan worden gesteld dat het model de genoemde parameters met een redelijke mate van nauwkeurigheid voorspelde en het wordt hierbij aanbevolen voor gebruik in Ile-Ife en andere delen van het stroomgebied van de Ogun-Osun rivier en in Nigeria.

Hoewel land, water en economische productiviteit van het gewas hoger waren bij water besparende maatregelen onder regenafhankelijke omstandigheden, verhoogde bodembehandeling om water te besparen en regelmatig onderhoud de gemiddelde seizoengebonden productiekosten vergeleken met de conventionele praktijk. Hoge kosten van de productie kunnen de opbrengsten verkregen door telers van het gewas verminderen, behalve wanneer er verbetering is in de marktprijs. Daarom moet duurzame praktijk van de waterbeheer maatregelen gepaard gaan met lagere productiekosten. Onder irrigatie condities worden de land en water productiviteit lager dan bij regenafhankelijke teelt. De productiviteit in het droge seizoen vermindert in relatie tot de ernst van de water tekorten. Gemiddelde gewasproductiviteit, water en economische water productiviteit bij alle zes waterbesparende maatregelen in het natte seizoen waren hoger dan bij volledige irrigatie in het droge seizoen. De productiekosten van het gewas waren in het droge seizoen aanzienlijk hoger dan de kosten onder de regenafhankelijke omstandigheden. De hogere water productiviteit onder regenafhankelijke omstandigheden in dit onderzoek is in overeenstemming met de bevinding dat in een aanzienlijk deel van de minst ontwikkelde landen en opkomende economieën er in vergelijking met geïrrigeerde landbouw een grotere kans is voor het verbeteren van de productiviteit van het water onder regenafhankelijke omstandigheden.

Uitbreiding van landbouwgrond in Ile-Ife kan vanwege de enorme benodigde investeringen niet haalbaar zijn. Daarom zou de aandacht voor de inspanningen om de voedselproductie in het gebied uit te breiden gericht moeten zijn op het verhogen van de productiviteit op de bestaande landbouwgronden en het verbeteren van productie-effectiviteit, resultaten die alleen kunnen worden bereikt door verbeterde variëteiten te gebruiken, in combinatie met verbeterde landbouwkundige praktijken. Landbouwkundige praktijken, vooral onder regenafhankelijke omstandigheden, zouden moeten worden ontworpen om de water productiviteit te verbeteren. Het verbeteren van de water productiviteit vereist verschuiving van vocht (transfer) waarbij bodem fysische omstandigheden, bodemvruchtbaarheid, gewas variëteiten en agronomie in combinatie worden gebruikt en beheerd om verdamping te verschuiven naar nuttige gewasverdamping. Tijdens de droge tijd zou het gewas moeten worden geïrrigeerd om maximale land en water productiviteit te realiseren. Het overslaan van irrigatie tijdens zaadvorming zou moeten worden vermeden om aanzienlijke vermindering van de opbrengst te voorkomen. Irrigatie bij het begin van afrijpen nadat de zaadvorming heeft plaatsgevonden kan worden overgeslagen. Onder water beperkende omstandigheden kan de bespaarde hoeveelheid water door het overslaan van irrigatie tijdens de bloei, vorming van peulen, zaadvorming en rijpheid worden gebruikt voor de teelt van andere gewassen en daarmee de alternatieve kosten verhogen. Incidentele regenval tijdens het droge seizoen zou moeten worden ingezet om irrigatiewater productiviteit van het gewas te verhogen.

Appendix R. About the author

Omotayo Babawande Adeboye has Bachelor of Engineering (B. Eng) degree in Agricultural Engineering in 2001 from the Federal University of Technology, Akure, Nigeria. He successfully completed a Masters of Engineering (M. Eng) degree in Soil and Water Engineering in 2005 at Federal University of Technology, Akure, Nigeria. He had industrial experiences at the Hydro-Construction and Engineering Company Nigeria Limited, Apapa, Lagos Nigeria. He worked as lecturer in the Department of Agricultural Engineering, Auchi Polytechnics, Edo State in Nigeria. Currently, he is an academic staff member in the Department of Agricultural and Environmental Engineering, Obafemi Awolowo University, Ile-Ife. He is a member of the Nigerian Institution of Agricultural Engineers (NIAE), Nigerian Society of Engineers (NSE) and registered with the Council for the Regulation of Engineering in Nigeria (COREN). He has published his research findings in both local and international journals. He is married and blessed with a daughter, Temitope Adeboye. He was offered a financial assistance under the Netherlands Fellowship Program to undertake his doctoral degree programme at UNESCO-IHE under so-called Sandwich construction.

Netherlands Research School for the
Socio-Economic and Natural Sciences of the Environment

D I P L O M A

For specialised PhD training

The Netherlands Research School for the
Socio-Economic and Natural Sciences of the Environment
(SENSE) declares that

Adeboye Omotayo Babawande

born on 27 December 1976 in Tonkere, Nigeria

has successfully fulfilled all requirements of the
Educational Programme of SENSE.

Delft, 28 May 2015

the Chairman of the SENSE board

Prof. dr. Huub Rijnaarts

the SENSE Director of Education

Dr. Ad van Dommelen

The SENSE Research School declares that Mr Adeboye Omotayo Babawande has successfully fulfilled all requirements of the Educational PhD Programme of SENSE with a work load of 34 EC, including the following activities:

SENSE PhD Courses

o Environmental research in context (2010)
o Research in context activity: Organising seminar and on farm training on 'drip irrigation systems for dry season farming in Ogun-Osun River Basin', Nigeria (2014)
o SENSE Writing Week (2014)
o Human induced soil degradation (2014)

Other PhD and Advanced MSc Courses

o World History of Water Management, UNESCO-IHE Delft (2010)
o Water Management System and Agronomy I, UNESCO-IHE Delft (2010)
o Aspects of Irrigation and Drainage, UNESCO-IHE Delft (2010)
o Summer school 'Climate Change in Integrated Water Management', UNESCO-IHE Delft (2011)
o Training course 'Improved drought early warning and forecasting to strengthen preparedness and adaptation to droughts in Africa', DEWFORA (2013)
o Training course 'Drought hazards, vulnerability and risk analysis tools in Africa', DEWFORA (2013)
o Multivariate Analysis, Wageningen University (2014)
o Introduction to R for statistical analysis, Wageningen University (2014)

Oral Presentations

o *Ensuring Food Security Through Productive And Sustainable Use of Land And Water in Nigeria.* 21st International Congress on Irrigation and Drainage: Water productivity towards food Security - International Commission on Irrigation and Drainage (ICID), 19-23 October 2011, Tehran, Iran
o *Biomass, yield and water productivity of Soybean under rainfed system in the sub-humid tropical climate of Nigeria.* 1st World irrigation Forum - International Commission on Irrigation and Drainage (ICID), 29 September-5 October 2013, Mardin, Turkey

SENSE Coordinator PhD Education

Dr. ing. Monique Gulickx

T - #0651 - 101024 - C0 - 240/170/14 - PB - 9781138028418 - Gloss Lamination